THE GEOGRAPHY OF SMALL FIRM INNOVATION

INTERNATIONAL STUDIES IN ENTREPRENEURSHIP

Series Editors:
Zoltan J. Acs
University of Baltimore
Baltimore, Maryland USA

David B. Audretsch
Indiana University
Bloomington, Indiana USA

The Geography
of Small Firm Innovation

by

Grant Black
Andrew Young School of Policy Studies
Georgia State University, Atlanta, Georgia

Kluwer Academic Publishers
Boston/ New York/Dordrecht

Library of Congress Cataloging-in-Publication Data

A C.I.P. Catalogue record for this book is available
from the Library of Congress.

ISBN 0-387-24184-1 (SC)   ISBN 1-4020-7612-6 (HC)   Printed on acid-free paper.

First Softcover printing, 2005

Printed in the United States of America.

9 8 7 6 5 4 3 2 1   SPIN 11370802

springeronline.com

For my parents, Carrol and Lorene Black

CONTENTS

FIGURES

TABLES

FOREWORD

It has long been recognized that advances in science contribute to economic growth. While it is one thing to argue that such a relationship exists, it is quite another to establish the extent to which knowledge spills over within and between sectors of the economy. Such a research agenda faces numerous challenges. Not only must one seek measures of inputs, but a measure of output is needed as well to estimate the knowledge production function. The identification of such a measure was a compelling goal for Zvi Griliches, if not the holy grail: "The dream of getting hold of an output indicator of inventive activity is one of the strong motivating forces for economic research in this area." (Griliches 1990, p. 1669).

Jaffe (1989) made a significant contribution to estimating the knowledge production function when he established a relationship between patent activity and R&D activity at the state level. Feldman and coauthors (1994a, 1994b) added considerably to this line of research, focusing on innovation counts as the dependent variable instead of patent counts. This work was particularly important given that many innovations are never patented. Feldman's work also differentiated by firm size and showed that knowledge spillovers from universities play a key role as sources of knowledge for small firms.

But much remains to be known concerning the knowledge production function and the role that spillovers play in the process. One critical area is the need for a measure of innovation that is more broadly based than patent counts. The innovation measure used by Feldman provided an answer, but the costliness of creating such a measure has meant that it has not been repeated. Another crucial area is the need to examine spillovers at a finer geographic area than that provided by the state. This is especially important given the tacit and sticky nature of knowledge.

Grant Black takes several important steps in correcting these deficiencies. First, he uses Small Business Innovative Research Phase II awards (SBIR) as a measure of innovation. Phase II awards are selective and have a higher rate of commercialization than do patents. Originally awarded in 1983, these awards have grown in number over time, amounting to over 4,000 awards for the period Black studies (1990-1995), with a value of over $2.5 billion in 1992 U.S. constant dollars. Second, Black focuses on the MSA as the unit of analysis, rather than the state. Knowledge production functions are estimated for five industries: chemicals and allied products, industrial machinery, electronics, instruments and research services.

In order to compare his results to those using a more traditional measure of innovation not restricted to small firms, Black estimates comparable MSA-based equations for patents in four of the five industries. For both measures of innovation, the methodology involves estimating a negative binomial hurdle model. In the first step, Black estimates whether a city innovates, using a zero-one dependent variable to indicate whether the MSA had received one ore more Phase II awards. The second step examines the rate at which the MSA innovates.

Black finds that geographic proximity matters for small-firm innovation. The relationship is particularly strong for *whether* a city receives one or more SBIR awards. The means are telling: 51% of the 137 cities that receive one or more SBIR Phase II awards have a research university while only 9% of the 136 cities that receive no awards have a research university. The mean number of industrial labs also varies by SBIR status. In SBIR cities, the mean number of labs is 67; in non-SBIR cities it is 5. Black finds that the *rate* at which cities innovate, as measured by the number of SBIR Phase II awards, also relates to university research activity but the university relationship is less strong than in the case of *whether* the city innovates. There is no evidence that the rate of innovation as measured by SBIR counts relates to the presence of industrial R&D labs.

Black finds patent activity to be much less concentrated among cities than is SBIR activity. To wit, 257 of the 273 cities received one or more utility patents during the interval studied. The presence of a research university plays less of a role in determining whether the city patents than it does in determining whether a city receives SBIR funding. The number of R&D labs, however, plays a comparable and significant role. When he focuses on the *rate* of patent activity, Black finds that university R&D expenditures affect the number of utility patents in three of the four industries studied. Reminiscent of his SBIR findings, no relationship is found between the number of industrial labs and the number of patents issued.

By introducing a new measure of innovation, and by shrinking the geographic unit of analysis, Black makes a considerable contribution to our understanding of the knowledge production function. An added plus is that Black does this with great clarity.

Paula Stephan
Andrew Young School of Policy Studies
Georgia State University
21 April 2003

ACKNOWLEDGEMENTS

This research owes its existence to Paula Stephan who has been a faithful mentor, and I am indebted to David Audretsch who envisioned this body of work in book form. I especially thank Zoltan Acs, David Audretsch, Adam Korobow, Sharon Levin, Paula Stephan, and Susan Walcott for in-depth reviews of this work, as well as Shiferaw Gurmu, Jerry Thursby, and Mary Beth Walker for insightful discussions. This work has also benefited from the comments of numerous colleagues, conference participants, and friends. I am grateful for the invaluable research assistance of Carrol Black and Albert Sumell. James Adams, Michael Page, and the U.S. Small Business Administration provided data without which this work would not have been possible. Lara Platt graciously provided editorial guidance in the production of this manuscript. Lastly, the success of this endeavor rests in many ways on the support of my family and friends who have been an unceasing encouragement, and so I dedicate this book to them.

1

INTRODUCTION

California's Silicon Valley, Massachusetts' Route 128, and North Carolina's Research Triangle conjure up images of intensely productive regions at the forefront of innovative activity. With these images in mind, politicians—particularly at the state and local level—increasingly are interested in growing their own regional hotspots of innovation.[1] One option that continues to gain momentum is the development of policies to attract and stimulate small business, given the increasing presence and importance of small businesses to economic activity in certain industries (Acs and Audretsch 1990, 1993; Acs, Audretsch and Feldman 1994; Pavitt et al. 1987; Phillips 1991). Effectively formulating such policies demands an understanding of the role that geographic proximity to knowledge plays in the innovation process of small firms. For instance, does it matter how closely a firm is located to similar firms, universities, R&D activity, or other resources? If this understanding is lacking or ignored, economic development and innovation policy can be misguided or altogether ineffective, especially with the temptation for policymakers to jump on the bandwagon of policy fads.

The importance of geographic proximity in innovation depends on the role of knowledge in the innovation process and the ease with which knowledge flows between agents and across space. Researchers such as Schumpeter (1942) and Arrow (1962) early on recognized the role of knowledge in production. They hypothesized that firms not only gained new knowledge through "learning by doing" but could also benefit from the knowledge generated by other economic agents, such as other firms or new employees. It is argued that these knowledge flows are influenced by the "knowledge infrastructure" located within a region. The knowledge infrastructure is comprised of formal and tacit knowledge embedded in institutions and individuals located in a region. "This infrastructure is of the greatest economic significance because industrial production is based ultimately on knowledge" (Smith 1997, p. 95). Indeed, the rapid expansion of technology and the growth of high-tech industries since the 1980s has caused some to speculate that knowledge will become the principle resource in the future (Bunk 1999).

Inroads into mapping the role of knowledge in the innovative process have been made but much remains to be examined, particularly the role that geography plays between knowledge and innovative activity. A growing literature has delved into understanding the effect that knowledge spillovers have in the development of technology, including the role science plays in knowledge creation. The New Growth Economics (Romer 1986, 1990) has established a theoretical link between knowledge and productivity. Technology is argued to be endogenously determined within the economy. Economic agents shape technological development through conscious actions such as R&D. This R&D affects the agents' productivity but also generates knowledge spillovers that increase the overall pool of knowledge available in the economy leading to economic growth beyond what would have occurred without the presence of the spillovers. This spilling over of knowledge is thus one mechanism through which ideas are transformed into innovation. Therefore, knowledge, which is seen as being produced, may spill over between agents in an economy and stimulate innovative activity.

Despite the widespread belief that knowledge is vital for economic growth, there is still no consensus on the ability to empirically identify the mechanisms through which knowledge is transferred. At one extreme, Krugman (1991a, 1991b) argues that it is futile to attempt to empirically explore spillovers—despite their likely existence— because they leave no identifiable trail. Others, more optimistic, claim spillovers can be traced through the documented transfer of knowledge and have applied a handful of measures to capture their effect, including citations to patents or publications, employment of scientific personnel, and R&D activity.[2]

Recent research has begun to search for empirical evidence of the existence of knowledge spillovers (Adams 1998; Griliches 1992). These efforts have predominately focused on knowledge-based industries where knowledge spillovers are believed to be more relevant to the innovation process. A subset of this literature has explored the role of geographic proximity in the spillover process (Almeida and Kogut 1998; Audretsch and Stephan 1996; Feldman 1994b; Jaffe et al. 1993; Saxenian 1985, 1996; Zucker et al. 1994). For example, Audretsch and Stephan (1996) and Zucker et al. (1994) find a relationship between the location of university scientists and biotechnology firms. Other evidence based on patent citations suggests that the concentration of knowledge within an area contributes to the clustering of innovative activity (Jaffe et al. 1993). These studies indicate that spillovers exist and that their localization is important to the innovation process.

The greatest hurdle in these attempts has been the lack of accurate, accessible measures of innovation and knowledge, and the inability to perform analyses at the appropriate spatial unit. Current measures are flawed by imperfect connections to innovation, aggregated units of observation particularly at the spatial level, or lack of data over time. Several studies (Feldman 1994a, 1994b; Jaffe 1989) have had to rely on state level data, though they recognize that urban centers would be the more appropriate level of analysis. Others (Acs et al. 1994; Anselin et al. 1997, 2000; Audretsch and Feldman 1996a, 1996b) have examined only one year of innovative activity because no time series of comparable data exists. Griliches (1990, p. 1669) articulates the urgency for improved data: "The dream of getting hold of an output indicator of inventive activity is one of the strong motivating forces for economic research in this area." Without new data sources the ability to more fully examine the role of knowledge spillovers in innovative activity is limited.

Beyond the limitations of current data, the body of evidence on knowledge spillovers remains incomplete. The vast majority of empirical research in this area has occurred within the last ten years.[3] Much of the empirical research has modeled spillovers within a knowledge production function framework first introduced by Griliches in 1979. A knowledge production function portrays a knowledge-based output as being produced by a pool of appropriated knowledge made up of various "knowledge inputs." Griliches constructed a function so that spillovers would be examined only in technological space. Griliches (1992) later acknowledged that this specification ignored the possibility of geographic spillovers raised by Huffman and Evenson (1991) regarding cross-state agricultural research spillovers.

To correct this shortcoming, a line of studies characterized by Jaffe (1989) and Feldman (1994a, 1994b) began to investigate whether spillover effects are geographically bounded. Many of the earlier studies examined spillovers at the state level due to data constraints. However, the state is generally considered too broad a region to effectively capture the intricacies expected in the spillover process. While the state as a unit of observation allows for an examination of how knowledge spillovers and agglomeration affect regional innovation, the substantial diversity of activity that exists within individual states cannot be captured at the state level. Yet, the creation and transfer of knowledge, as well as access to other resources, arguably best takes place in smaller geographic areas, such as cities (Lucas 1993). If knowledge is sticky, as von Hippel (1994) contends, so that the cost of transmitting knowledge rises with distance, firms locate near sources of knowledge to reduce costs. Firms, therefore, have an incentive to cluster in urban areas that facilitate the flow of ideas between individuals and firms (Glaeser 2000; Lucas 1998) This clustering of knowledge can stimulate innovation within these areas, while other cities—even within the same state—without such clustering may see little innovative activity. To correct this drawback, emphasis has shifted towards urban centers of economic activity, such as metropolitan areas (Anselin et al. 1997, 2001; Jaffe et al. 1993). The problem of limited data, however, has restricted efforts to isolate spillover effects at these smaller units of observation.

This book addresses the meager evidence of the role of knowledge spillovers in the innovation process and the measurement of innovative output for small firms. The research presented here introduces the Small Business Innovation Research

(SBIR) Program Phase II award as a novel indicator of innovative output for small high-tech firms in the United States.[4] The SBIR Phase II award, as a measure of innovation, provides consistent data since 1983 (the inception of the SBIR Program) and has qualities that allow it to serve as a proxy for actual innovation by small firms. The Phase II award is an intermediate output of research, similar to a patent, with a strong correlation to a final commercialized innovation. Moreover, because of the nature of the SBIR Program, the Phase II award is the only readily available measure of innovation focusing exclusively on small firms.

Using this unique measure in empirical analysis, this research examines the role geography plays in the innovative process at the metropolitan area level during the first half of the 1990s. Unlike most previous work that focuses on a one-year period, this study examines innovative activity aggregated across six years, which provides a view of the "average" strength of spillovers in a multi-year period. Of particular interest are (1) the role universities and industry play in providing knowledge spillovers and (2) the influence of agglomeration through the concentration of economic activity. The analysis spans 273 metropolitan areas, including those with and without measured innovative activity.[5] The econometric technique employed in this research (a negative binomial hurdle model) more accurately accounts for the distributional characteristics of innovation data than previous work that used count data, such as innovation or patent counts. Furthermore, this analysis distinguishes between the impact of spillovers on the presence of innovative activity and on the rate of innovation, unlike previous work (Anselin et al. 1997, 2000) that looked only at areas engaged in innovation.

The clustering of innovative activity in specific regions has been well documented (Oakey 1984; Saxenian 1985, 1996; U.S. Small Business Administration 2000). A growing desire exists to understand how these innovative centers developed and continue to grow. Since the 1970s, evidence continues to demonstrate that small firms can substantially contribute to innovation and overall economic growth (Acs 1999; Acs and Audretsch 1990; Acs, Audretsch and Feldman 1994; Korobow 2001; Phillips 1991). This is in part due to small firms' access to knowledge generated by sources outside the firm, such as universities and large, established firms. Therefore, small firms may be particularly impacted by the geographic boundaries of spillovers.

This research focuses on the innovation process for small, high-tech firms to develop a clearer picture of the specific effect of spillovers on small firms than the majority of past research on spillovers has provided. Identifying the links between knowledge sources and innovating small firms is important, particularly as the expansion of the high-tech sector has spawned the growth of small firms. The findings can influence policy to stimulate regional economic development through the small business sector. For instance, policies can be implemented that attract and strengthen knowledge sources in innovation-poor areas or that target the creation of small firms in areas with adequate knowledge infrastructures.

Consistent with previous research, the findings presented in this book indicate that geographic proximity matters to small-firm innovation (as measured by Phase II awards), though with varying degrees of significance across high-tech industries. The strength of that relationship, however, depends on how innovation is measured. Whether or not small firms in a metropolitan area innovate at all depends on the presence of external sources of knowledge, including industrial R&D labs and research- oriented universities, and the presence of agglomerative economies indicated by the concentration of industry-specific employment. The rate of innovation, however, depends less on these determinants, suggesting that small firms engaged in SBIR activity rely on external knowledge in the innovation process but that other factors—at least for SBIR activity—play a dominant role in determining the rate of innovation for these high-tech small firms. This is in part evident by the variation in the local technological infrastructure's effect on Phase II activity disaggregated by SBIR funding agency. The evidence suggests that agency effects influence the impact of the technological infrastructure on SBIR activity.

To provide a means of comparison to the analyses of SBIR activity among small firms, an analysis of patent activity at the metropolitan level is presented. This patent analysis refines earlier studies of patents at the state level (Jaffe 1986, 1989; Feldman 1994) and updates previous studies of innovative activity in metropolitan areas during the 1980s (Anselin et al. 1997, 2000; Feldman and Audretsch 1999; Jaffe et al. 1993; Varga 1998). The results identify a relationship between the local technological infrastructure and patents, which is consistent with this study's findings for SBIR Phase II activity as well as with previous research. Knowledge spillovers and agglomerative economies lead to increased patent activity. As with Phase II activity, university spillovers impact the likelihood and level of patent activity across industries, while spillovers from R&D labs have a significant impact only on the likelihood of patent activity. Variation in the effect of the local technological infrastructure also exists across industries, particularly for university spillovers and business services. In contrast to Phase II activity, agglomerative economies from the concentration of industry-level employment, availability of business services, and size of the metropolitan area play a significant role in patenting across industries.

The book is organized as follows. The first two chapters provide a conceptual background on the role of geography in the innovation process and an overview of the SBIR Program and measures of innovation. Chapter 2 describes the SBIR Program, presents the strengths of using the SBIR Phase II award as a measure of small-firm innovation, and examines the geographically skewed nature of innovative activity in the United States. Chapter 3 discusses how external economies arise from agglomeration and knowledge spillovers within a local technological infrastructure.

The remaining chapters present the empirical analyses. Chapter 4 develops an empirical model based on a knowledge production function to evaluate innovative activity at the metropolitan level during 1990-95. The chapter includes a description of the unique data set constructed for the empirical analyses, followed by a discussion of the econometric technique adopted to examine the likelihood and rate of innovative activity. Chapter 5 focuses on Phase II awards as the measure of innovative activity among high-tech small firms in the United States. Chapter 6 explores the existence of agency effects related to Phase II activity by disaggregating the analysis for Phase II awards by funding agency. Chapter 7 presents empirical findings for patents as the measure of innovative activity. Chapter 8 summarizes the empirical findings and focuses on the policy implications emanating from this body of work.

2

THE SMALL BUSINESS
INNOVATION RESEARCH PROGRAM

LEGISLATIVE DEVELOPMENT

Beginning in the mid-1970s, the federal government became increasingly concerned with the role of small business in federal R&D activities and in the national economy. The consensus among policymakers and politicians during this time was that small business was vital to economic activity but underrepresented in federal R&D activities. According to a 1981 U.S. Senate Report,

> Numerous studies have shown that small businesses are our Nation's most efficient and fertile source of innovations. Yet only 3.5 to 4 percent of the Federal R&D dollar is spent with small firms. This underutilization of small business in Federal R&D programs is especially regrettable when considering the highly successful track record of small firms in generating jobs, tax revenue, and other economic and social benefits (U.S. Congress 1981, pp. 4-5).

Concern for the prominence of the United States in the global economy simultaneously arose as the gap between the U.S. and other nations diminished, particularly in high technology sectors (National Science Board 1977; U.S. Department of Commerce 1977; U.S. General Accounting Office 1981).[6] Political efforts began in the late 1970s and 1980s to maintain and increase the competitive advantage of the U.S. (Carter 1979; Wessner 2000).

Armed with perceptions about the importance of small business and an impetus for stimulating U.S. competitiveness, the federal government turned its attention to research-oriented small businesses. An array of legislation was spawned that drew the federal government into a more active role in cooperative research and technology arrangements. Table 1 highlights major federal research and technology legislation enacted in the 1980s.

Table 1. Significant Federal Research and Technology Legislation in the 1980s

1980	• Stevenson-Wydler Technology Innovation Act
	• Bayh-Dole University and Small Business Patent Act
1982	• Small Business Innovation Development Act
1984	• National Cooperative Research Act
1986	• Federal Technology Transfer Act
1988	• Omnibus Trade and Competitiveness Act
1989	• National Competitiveness Technology Transfer Act

Source: Wessner, Charles, ed. *The Small Business Innovation Research Program: An Assessment of the Department of Defense Fast Track Initiative*, Washington, DC: National Academy Press, 2000, pp. 18-19.

Three of these acts have direct implications for the small business sector. The Bayh-Dole Act in part allows grantees and contractors—who are small firms, universities, or non-profit organizations—to retain title to inventions arising from federally funded R&D. The Omnibus Trade and Competitiveness Act established the Advanced Technology Program, which targets research in specific technologies and funds many small firms. The Small Business Innovation Development Act created the federal SBIR Program, which strives to increase small-firm innovation and commercialization.

The National Science Foundation (NSF) instituted the precursor to the federal Small Business Innovation Research Program in 1977. The purpose of this initiative was "to increase the opportunity of small high tech companies to participate in NSF research and to stimulate the conversion of the research results into technological innovation and commercial applications for their potential economic benefits to the nation" (Tibbetts 1996 p. 1). NSF initially designated $5 million towards its SBIR program. In 1979 President Carter proposed a $10 million increase to the NSF SBIR program given its supposed success in generating quality research with potentially significant market and social returns.

The U.S. Congress expanded the Small Business Innovation Research program largely in response to the perceived success of the NSF program. A federal SBIR program was created in 1982 under the Small Business Innovation Research Act as

an R&D policy targeting small businesses.   The Congress reauthorized the
program's continuance in 1986, 1992, and 2000.   The initial goals of the SBIR
Program, explicitly stated in the legislation, were to:

1. stimulate technological innovation;
2. use small business to meet federal R&D needs;
3. foster and encourage participation by minority and disadvantaged persons in technological innovation; and
4. increase private-sector commercialization of innovations derived from federal R&D (U.S. Congress 1982).

Noticeable changes were implemented in 1992 under the Small Business
Innovation Research Program Reauthorization Act, including revised funding
guidelines, legislative emphasis on the program's commercialization goals, and
creation of a sister program—the Small Business Technology Transfer (STTR)
Program. Over time political interest in the SBIR Program had shifted increasingly
towards the potential economic impact of commercialization of small-firm
innovations.   This prompted the Congress to reprioritize its goals for the SBIR
Program.   Four criteria were imposed in the 1992 reauthorization legislation as
guidelines for evaluating the commercial potential of an SBIR Phase II proposal:

1. The firm's record of successfully commercializing prior SBIR or other research;
2. Evidence of funding commitments from private sector or non-SBIR public funding sources;
3. Evidence of post-SBIR commitments for the commercialization development of the proposed research; and
4. The presence of other indicators of the commercial potential of the proposed research…. (U.S. Congress 1992)

These criteria were explicitly stated to more clearly direct SBIR research reviewers
and decision makers in the SBIR selection process by outlining desired
characteristics of proposed research.   They also provided firms interested in
pursuing SBIR funding further insight on what contributes to a successful SBIR
proposal, allowing firms to better evaluate the submission decision and better target
proposals.
    Apart from the NSF pilot program, the federal SBIR Program arose from the
growing literature citing the significant contribution of small firms to economic
growth through innovations and job creation[7] and suggesting that small firms face
disproportionately higher costs in financing R&D than large firms, particularly in
the early stages of R&D (Hubbard 1998; Jewkes 1969; Stiglitz and Weiss 1981).
Concurrently, evidence indicated the lack of federal R&D funds captured by the
small business sector (Cardenas 1981; National Science Board 1977; U.S. Office of
Management and Budget 1977; Zerbe 1976).[8]   At a time when the Congress was
actively involved in policies aimed at stimulating the economy, this evidence
prompted policymakers to target small business in its economic initiatives.   The

expansion of the NSF SBIR program to a government wide program was a major component in efforts to more effectively capture the economic benefits emanating from the small business sector.[9]

SBIR PROGRAM STRUCTURE

The structure of the federal SBIR program followed the design of the NSF pilot program but expanded its coverage across all applicable government agencies. The Small Business Administration (SBA) was charged with overseeing the SBIR Program. The SBA's role includes issuing general policy directives regarding the SBIR Program, monitoring the SBIR activities of participating agencies, evaluating the program, and increasing firms' awareness of (and commercialization success in) the program.

Federal Agency Participation

Participation in the SBIR Program is mandatory for all federal agencies with annual external R&D budgets in excess of $100 million. Each agency is independently responsible for its solicitation of SBIR research proposals that fall within the boundaries of the agency's mission, the competitive selection of projects based on each agency's standard review practices, and the disbursement of its SBIR funds. Agencies independently implement review procedures, which typically follow an agency's other funding guidelines. There are fundamentally two different approaches by which agencies evaluate proposals: "line review" and "peer review" (Busch 1999). Under a line review system, an agency's line management personnel perform evaluation; this is the method used by both the Department of Defense (DOD) and the National Aeronautics and Space Administration (NASA) for SBIR review. Peer review relies on independent third party evaluation. SBIR proposals at the National Institutes of Health (NIH) and the National Science Foundation (NSF), for instance, face peer review as do their other external funding programs; the Department of Agriculture also uses peer review to evaluate SBIR proposals.

Table 2 lists the ten agencies that currently participate in the Program and reports the amount of SBIR funds distributed by agency from 1990 to 1995. Given the Congressional mandate, agencies with the largest budgets have the largest SBIR programs. The distribution of SBIR finances (and therefore also the number of awards) is thus highly dependent on the size of the agencies' external R&D budgets and highly skewed given the skewed funding of agencies. The Department of Defense is by far the largest agency in the SBIR Program both in terms of the number of awards and the amount of funds disbursed, providing over $1.3 billion for SBIR awards between 1990 and 1995—or more than 40 percent of the value of all SBIR awards during this period. The next largest participant is the Department of

Health and Human Services, where contributions have risen dramatically in recent years as the National Institutes of Health budget has grown. The National Aeronautics and Space Administration is next, followed by the Department of Energy and the National Science Foundation. Taken together, these five agencies comprised more than 95 percent of the value of all SBIR awards from 1990 to 1995. The remaining five agencies—the Departments of Agriculture, Commerce, Education and Transportation and the Environmental Protection Agency—accounted for just over 4 percent of SBIR funds.

Table 2. Federal Agencies Participating in the SBIR Program[*]

Agency	SBIR Funds Distributed, 1990-95 (millions of dollars)
Department of Agriculture	40.7
Department of Commerce	17.5
Department of Defense	1,366.8
Department of Education	16.4
Department of Energy	296.8
Department of Health and Human Services	762.9
Department of Transportation	36.8
Environmental Protection Agency	27.0
National Aeronautics and Space	524.7
Administration	223.1
National Science Foundation	
	3,312.7
Total	

*The Nuclear Regulatory Commission also participated in the SBIR Program in 1990-95 but does not currently participate due to cuts dropping its budget below the $100 million minimum requirement.

SBIR funding is disbursed either through contracts or grants based on each agency's procedures for funding external research. Not surprisingly, the three agencies that rely on peer review to evaluate research proposals rely on grants. The other agency using grants is the Department of Energy. By contrast, those that rely on line review issue contracts—including the Departments of Commerce, Defense, Education, and Transportation, the Environmental Protection Agency, and the National Aeronautics and Space Administration.

It is important to realize that the SBIR Program does not provide *new* funds for R&D but *redistributes* a portion of existing R&D funds to small firms. Participating agencies are required by SBIR legislation to set aside a mandated fraction (2.5 percent since 1997) of their external R&D budgets for the SBIR Program. In the early stages of the program, the set-aside percentage was phased in to a standard level, and a two-tier system was enacted to ease the transition of the SBIR set-aside for agencies with extremely large R&D budgets. The standard set-aside percentage rose in the first four years of the program, from an initial 0.2 percent in 1982 to 1.5 percent in 1986. For agencies with external R&D budgets exceeding $10 billion, the initial rate was 0.1 percent, rising to 1.25 percent by the fifth year of the program. The 1992 reauthorization legislation eliminated the two-tier set-aside scheme and raised the set-aside for all participating agencies to 1.5 percent in 1993, 2.0 percent in 1995, and 2.5 percent in 1997.

Eligibility and Incentives for Firm Participation

As outlined in its defining legislation, the intention of the SBIR Program is to increase the economic benefits generated by R&D and innovation at research-oriented small firms and to target small firms for participation in federally funded R&D and government procurement. The program, therefore, provides federal funding only to small firms as an incentive mechanism for small firms in high-technology sectors to increase innovation and commercialization. The SBA has established guidelines on the eligibility of firms for participation in the SBIR Program (U.S. Congress 1982). To receive SBIR funding, firms must be independently owned and operated, be for-profit, employ no more than 500 employees (including employees of subsidiaries and affiliates), and be predominately owned (at least 51 percent) by U.S. citizens or legal permanent residents. An eligible firm cannot be construed as a dominant firm within the field in which it proposes to perform SBIR projects. Moreover, the principal investigator for an SBIR project must be principally employed by the small firm during the project, and a majority of the research for the SBIR project must occur at the firm (though a portion can be subcontracted outside the firm).

In addition to subsidizing private R&D, the SBIR Program offers small firms other economic incentives to pursue SBIR research. As previously mentioned, an explicit objective of the SBIR Program is to increase the use of small firms in meeting government R&D needs. Firms that successfully complete SBIR research, therefore, have a potential market through the funding agency for their research outcomes. Surveys in the 1990s by the U.S. General Accounting Office (GAO) and the Department of Defense of firms awarded SBIR funds indicate that between 35 percent and almost 53 percent of sales attributed to SBIR projects came from the federal government (U.S. General Accounting Office 1998). SBIR legislation also allows firms to seek patentable intellectual property rights for outcomes from SBIR research. Firms may also retain government property used in SBIR research for at least two years from the beginning of Phase III.

Multi-Stage Design

The SBIR Program was designed as a multi-stage program to help maintain the quality of research throughout the innovation process that is supported by federal funding. The argument is that agencies can more easily monitor and evaluate research progress by imposing "short" funding cycles. The SBIR Program specifically consists of three sequential phases. The first phase (Phase I) is a competitive awarding of limited federal research funds for the short-term investigation of the scientific merit and feasibility of a research proposal. Phase I is designed to determine the research capabilities of a firm and the promise of a particular project. SBIR guidelines cap Phase I funding at $100,000 per award and allow up to six months for Phase I completion.[10] Competition for Phase I awards is high, with approximately 12 to14 percent of proposals successfully being granted a Phase I award (Tibbetts 1998).

Phase II is a competitive awarding of additional federal funds to continue research and begin development of the idea pursued in Phase I. Eligibility for Phase II funding, therefore, is restricted to research that has successfully completed Phase I. A Phase II award has a maximum level of $750,000 and typically is restricted to two years.[11] Approximately 40 percent of Phase I award recipients receive Phase II funding so that only 5 percent of SBIR proposals achieve Phase II (Tibbetts 1998). Selection for Phase II emphasizes research projects with not only strong scientific merit but also strong commercial potential. As described earlier, this emphasis on commercialization became more pronounced in the 1992 SBIR reauthorization, which amended the mandated criteria for Phase II selection.

The third phase (Phase III) is devoted to product development and commercialization arising from Phase II projects. Phase III is ineligible for SBIR funds. Firms at this stage are required to raise private-sector financing. A federal agency, however, can provide funds in Phase III if that agency expects to purchase the Phase III results, but the agency must finance Phase III using non-SBIR funds. The Small Business Administration aids Phase III commercialization in part through the Commercialization Matching System, a database with information on all SBIR awards and potential private investors. The SBA can then match SBIR projects with possible investors based on technological, industrial, and geographic preferences and investment thresholds. Several agencies also offer assistance in commercialization efforts. The Department of Energy, for example, provides individualized assistance to Phase II award recipients in locating and approaching potential private-sector investors through its Commercialization Assistance Program.

An on-going concern with the SBIR Program is the length of time that elapses between the completion of Phase I and the awarding of Phase II funding (Audretsch, Link and Scott 2000). The negative impact most often associated with this gap is the suspended funding during the gap. Many argue—particularly from the recipient firm perspective—that research projects cannot adequately progress between Phase I and Phase II without continuous funding. The decline or absence of research funding can disturb a firm's ability to maintain sufficient research staff, facilities, and equipment thereby increasing the firm's time-to-market (Cahill 2000). The time involved in commercialization is particularly important to high-tech firms in markets where the rate of technological obsolescence can be rapid (U.S. General Accounting Office 1998).

Several agencies have initiated policies aimed at reducing or eliminating the potential negative impacts associated with this funding gap. For example, DOD instituted its Fast Track program in 1996 to provide continuous funding between Phase I and Phase II, and NIH followed with its own version of Fast Track.[12] These Fast Track programs target firms with rapid commercial potential, allowing firms to simultaneously submit Phase I and Phase II proposals. Eligibility for Fast Track requires firms to demonstrate early-stage commitment or serious intent by outside private investors to be involved in the proposed research project.

While these Fast Track programs reduce the funding gap for some types of firms, they are not applicable to all firms participating in the SBIR Program (Archibald and Finifter 2000). One reason is that SBIR research performed by many firms is not at a stage in which both Phase I and Phase II proposals can be simultaneously developed. Therefore, the funding gap remains for many firms. To diminish the gap for this type of firm, the Department of Energy (DOE), for instance, allows Phase I recipients to submit Phase II proposals before completion of Phase I in order to speed up the application process.

Size of the SBIR Program

The SBIR Program has become the largest federal R&D program for small business. Over 45,000 awards worth over $8 billion (in 1998 dollars) have been granted through the SBIR Program since 1983 (Wessner 2000). Funds disbursed for the SBIR Program have reached over $1 billion annually since 1998.

The SBIR Program has grown dramatically since its inception, as can be seen from Table 3, which shows the number and value (in millions of constant 1992 dollars) of SBIR awards from 1983 to 1995. Phase I awards have grown on average approximately 13.5 percent annually, but the rate of growth changed dramatically over time. In the first four years of the program, Phase I awards grew between 40 to 45 percent; since 1988 they have grown by 10 percent or less. In two years (1988 and 1995) during this period the number of Phase I awards actually fell from the previous year's level. The total number of awards, however, fell only once (in 1994) over the thirteen-year period. Phase II awards have grown at an average annual rate of over 14 percent.[13] The number of Phase II awards dropped by almost 20 percent between 1993 and 1994 but rebounded quickly in the following year by nearly 36 percent.

The value of SBIR awards has also increased considerably from 1983 to 1995, driven by increases both in the number of funded projects and in the budgetary set-aside requirements during this same period. In its first year (which included only Phase I awards) just over $60 million was given to small firms for R&D. In the second year more than $142 million was awarded with almost $80 million designated for Phase II projects. The total value of Phase I awards was more than four times higher in 1995 than in 1983, while the total value of Phase II awards increased over eight fold.

Table 3. Number and Dollar Value of SBIR Awards, 1983-95

Fiscal Year	Phase I Awards		Phase II Awards		Total Awards	
	Number	Value[1]	Number	Value[1]	Number	Value[1,2]
1983	686	60.8	--	--	686	60.8
1984	999	63.2	338	79.6	1,337	142.8
1985	1,397	88.0	407	165.5	1,804	253.5
1986	1,945	122.2	564	247.4	2,509	369.7
1987	2,189	132.0	768	290.0	2,957	422.0
1988	2,013	118.4	711	289.1	2,724	451.9
1989	2,137	120.0	749	358.6	2,886	481.4
1990	2,346	126.2	837	365.2	3,183	492.2
1991	2,553	131.4	788	345.1	3,341	496.4
1992	2,559	127.9	916	371.2	3,475	508.4
1993	2,898	150.0	1,141	478.1	4,039	680.1
1994	3,102	209.7	928	450.7	4,030	682.9
1995	3,085	250.2	1,263	648.6	4,348	898.8
1983-95	27,909	1,700.0	9,410	4,089.1	37,319	5,940.9

Source: Small Business Administration. Office of Technology. *Small Business Innovation Research (SBIR) Program Annual Report 1995.* Washington, DC: U.S. Government Printing Office, 1997.
[1]In millions of constant 1992 dollars
[2]Summing values for Phase I and Phase II will not equal reported totals for 1988-95 because the totals include award modifications.

GEOGRAPHIC DISTRIBUTION

Geographically, the distribution of SBIR awards is highly concentrated, with firms in a handful of states and metropolitan areas receiving the vast majority of SBIR financing. This pattern is not unique to SBIR awards.[14] Regardless of the measure, innovative activity is predominately concentrated on the east and west coasts, with pockets of activity scattered in the interior. By way of example, Figure

1 shows the distribution of total R&D expenditures (in constant 1992 dollars) by state for 1990-95. Top-ranked California had approximately three times more R&D expenditures than the next highest state, New York. Four of the top five R&D states are on the east or west coast, with Michigan the only exception.[15] R&D activity on the east coast is concentrated in the northeast in New Jersey, New York and Massachusetts. Moreover, R&D expenditures are concentrated in less than a quarter of all states. Twelve states had more than $20 billion of R&D expenditures, while 27 states—more than half of all states—had less than $10 billion of R&D expenditures. The majority of these low-level R&D states are located in interior regions of the United States, such as the Midwest.

Another example of the highly concentrated nature of innovation is provided by Figure 2, which depicts the distribution of utility patents by state for 1990-95.[16] The distribution resembles that for total R&D expenditures during the same period, although concentration on the east and west coasts is less pronounced. Yet, California, New York and New Jersey remained in the top five states in terms of number of patents granted, as they also were in terms of total R&D expenditures. Texas (third) and Illinois (fifth) joined the ranks of the top five states, while Michigan was ranked sixth in number of patents. The concentration of patent activity is also skewed to relatively few states, as are R&D expenditures. Over 316,000 utility patents were granted in the United States in 1990-95. Ten states received more than 10,000 patents, accounting for almost two-thirds of all utility patents in the U.S. The majority of states made up the remaining third. Thirty-one states had less than 5,000 patents, and almost half of these states received less than 1,000 patents.

The skewed distribution of patents is even more striking when seen at the metropolitan area level. Of 273 metropolitan areas in the United States, 225 received less than 1,000 patents in 1990-95. Only 14 metro areas received more than 5,000 patents, with half of these receiving over 10,000 patents (USPTO 1998). Table 4 lists the top five metropolitan areas in 1990-95 for number of utility patents received. All but one (Chicago) of the top five metropolitan areas receiving patents were in either California or the Northeast. While California had the greatest number of statewide patents, the New York metropolitan area received the most number of patents at the metro area level; San Francisco and Los Angeles were ranked second and third, respectively. The top five metro areas combined accounted for almost a third of all patents in 1990-95, further indication of the geographically skewed nature of patent activity.

SBIR activity follows a similar geographic pattern to that of R&D expenditures or patents. In the only in-depth analysis of the geographic distribution of SBIR activity, Tibbetts (1998) finds that the top one-third of U.S. states accounted for almost 85 percent of all SBIR awards from 1983-96, while the bottom third of states received just over 2 percent. The disparity across states during this period was enormous; California, the state receiving the most SBIR awards, received over 9,000 while the lowest ranked state, Wyoming, had only 11 in the same fourteen-year

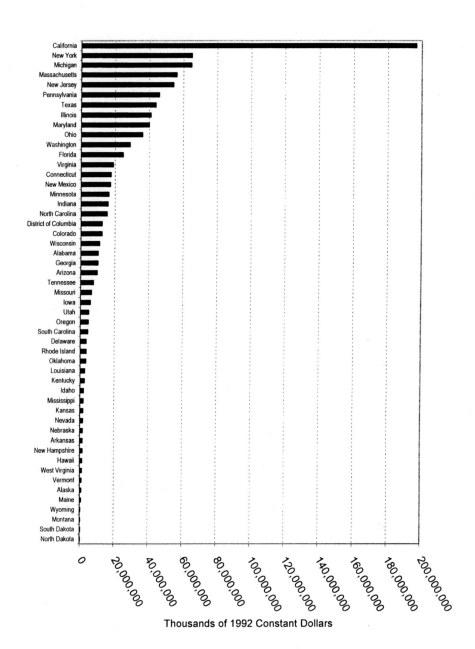

Source: National Science Foundation/Division of Science Resources Statistics.

Figure 1. Total R&D Expenditures by State, 1990-95

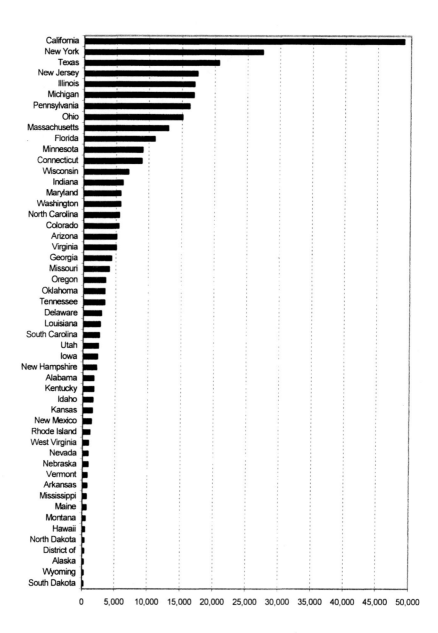

Source: USPTO. "United States Patent Counts by State, County and Metropolitan Area (Utility Patents 1990-1999)," April 2000.

Figure 2. Distribution of Utility Patents by State, 1990-95

Table 4. Top Five Metropolitan Areas Receiving Utility Patents, 1990-95

Metropolitan Area (Number of Patents)
New York
(33,212)
San Francisco
(20,271)
Los Angeles
(19,255)
Chicago
(14,092)
Boston
(13,590)
Percent of all Utility Patents Received by the Top Five Metro Areas
31.7%

Source: USPTO. "United States Patent Grants by State, County, and Metropolitan Area (Utility Patents 1990-1999)," April 2000.

period. Tibbetts compares state SBIR activity to other high-tech and R&D activities within a state. He finds that 80-85 percent of total R&D expenditures, venture capital investment, and the number of high-tech small firms were located in the same 17 states that received the most SBIR awards.

Figure 3 shows the distribution of Phase II awards in the first half of the 1990s across the United States. The stark differences in awards across states are striking and reflect Tibbetts' findings. A few states receive high numbers of awards, while several states receive almost no awards. States with little Phase II activity tend to be in the Midwestern region and Southern beltway. The highest concentration occurs along the east coast and in California, with considerable SBIR activity also in interior pockets such as Colorado and Texas.

Table 5 narrows the geographic focus of SBIR activity to the metropolitan area level. The table lists the top five metropolitan areas by number of Phase II awards received from 1990-95 in five broad industries that encompass the high-technology sector. At the metropolitan level, SBIR activity is even more concentrated than patenting in the United States. Coastal regions dominate as with patents and R&D expenditures. The majority of SBIR activity occurs in a small number of metro areas, with many metropolitan areas having no SBIR activity at all. Five metro areas account for approximately 50 percent or more of all Phase II awards across these five industries. Boston is ranked either first or second across all five industries, with San Francisco and New York among the top five in every industry as well. Denver is

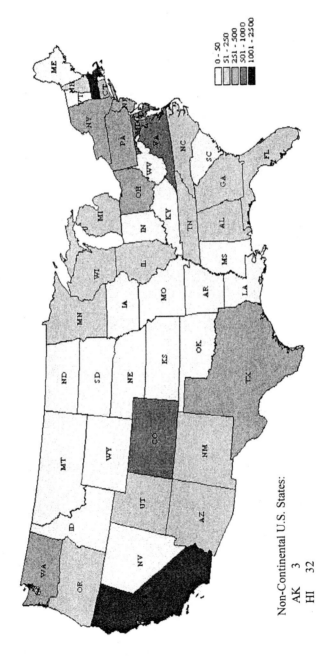

Non-Continental U.S. States:

AK   3
HI   32

Source: U.S. Small Business Administration

Figure 3. Number of SBIR Phase II Awards by State, 1990-95

Table 5. Top Five Metro Areas Receiving Phase II Awards by Industry, 1990-95
(Number of Awards)

Chemicals & Allied Products (SIC 28)	Industrial Machinery (SIC 35)	Electronics (SIC 36)	Instruments (SIC 38)	Research Services (SIC 87)
San Francisco (51)	New York (46)	Boston (132)	Boston (138)	Boston (371)
Boston (47)	Boston (28)	San Francisco (95)	Los Angeles (136)	Washington DC (212)
New York (44)	Seattle (27)	New York (76)	San Francisco (78)	Los Angeles (164)
Denver (29)	San Francisco (24)	Los Angeles (71)	Washington DC (71)	San Francisco (142)
Washington DC (29)	Lancaster, PA (16)	Washington DC (47)	New York (65)	New York (125)

Percent of all Phase II Awards Received by the Top Five Metro Areas				
52.6%	49.3%	57.6%	51.0%	56.8%

Table 6. Top Five States and Metropolitan Areas for Innovation, 1990-95[1]

	R&D Expenditures	Utility Patents	Phase II Awards
Top Five States	California	California	California
	New York	New York	Massachusetts
	Michigan	Texas	New York
	Massachusetts	New Jersey	Virginia
	New Jersey	Illinois	Maryland
Top Five Metro Areas	NA	New York	Boston
		San Francisco	San Francisco
		Los Angeles	Los Angeles
		Chicago	Washington, DC
		Boston	New York

[1]The ranking of Phase II Awards is based on awards received by firms in high-technology industries, as defined by the five industries for which data is reported in Table 2-5.
NAData for R&D expenditures are unavailable at the metropolitan area level.

ranked only in chemicals (fourth) and Lancaster, Pennsylvania, only in machinery (fifth).

It has been shown that R&D expenditures, patents, and SBIR Phase II awards demonstrate a strong tendency to be geographically concentrated in the United States. To summarize this common pattern of geographic concentration, Table 6 compares the top five states and metropolitan areas for R&D expenditures, utility patents, and Phase II awards. As previously highlighted, innovative activity occurs predominately in California and the Northeast. California is ranked first in all three measures. New York is among the top five states in all three measures. Massachusetts is in the top states for R&D expenditures and Phase II awards, while New Jersey ranks in the top states for R&D expenditures and patents. Innovation as measured by patents shows the most dispersed pattern among the top five states, with two of the five being central states (Texas and Illinois). At the metropolitan level, four of the top five areas for patents and Phase II awards are the same and are located in California, New York, or Massachusetts. The metro areas that are outliers are Chicago, ranked fourth for patents, and Washington, DC, ranked fourth for Phase II awards. The high concentration of SBIR activity in Washington, DC, is likely due to the proximity of firms to government agencies, particularly DOD and HHS.

The skewed geographic distribution of SBIR awards has raised concern about the program's effectiveness in achieving its goal of stimulating innovation and commercialization throughout the United States (U.S. General Accounting Office 1999). The 1992 reauthorization legislation implicitly addressed this issue when concern was raised about potential differences in local awareness of the SBIR Program. Increased efforts have been undertaken to heighten awareness and improve dissemination of information about the program. As a consequence, agencies regularly host informational seminars to educate small, high-tech firms about their SBIR programs. Beginning in 1999 the SBA started efforts to increase SBIR activity in states with low SBIR participation (SBA 1999).[17] The twenty-three targeted states are:

Alaska	Louisiana	Oklahoma
Arkansas	Maine	Rhode Island
Delaware	Mississippi	South Carolina
Hawaii	Missouri	South Dakota
Idaho	Montana	Vermont
Indiana	Nebraska	West Virginia
Iowa	Nevada	Wyoming
Kentucky	North Dakota	

The goal of the program is to assist states in increasing SBIR participation among local firms through SBIR-related information dissemination, training, and support.

Proposed legislation for the 2000 reauthorization explicitly stated concern about the geographic inequalities and called for the development of the Federal and State Technology Partnership program (State Science and Technology Institute 2000). The proposed program would provide matching funds from a $10 million start-up budget for states to develop their own SBIR initiatives to improve participation among local firms. The proposed program would not restrict eligibility to low-participation states but would provide a greater matching ratio for these states. State efforts would likely focus on initiatives such as pre-Phase I project organization, bridge funding, networking, and commercialization.

Many state technology offices already portray the SBIR Program as a valuable option for early-stage R&D funding to small firms within their state (Busch 1999; Engert 1998). Several states have created state-level SBIR programs to stimulate high-technology small business within their state based on the belief that SBIR activity acts as a significant incentive for high-tech business activity. These initiatives tend to occur in states with less developed technology sectors, such as Kansas or Wyoming. A major goal of these programs is to systematically increase SBIR awareness and encourage participation among small firms, in the hope that SBIR activity will contribute positive economic gains at the state level (Office of Research 2000; The Illinois Coalition 2000).

A MEASURE OF INNOVATIVE ACTIVITY

Measures of innovative activity rely almost exclusively on proxies drawn from innovative inputs, such as research and development (R&D) expenditures and employment, or intermediary innovation outputs, like patents. These measures, however, are not clearly linked to innovation. High levels of R&D expenditures, for example, may not coincide with large numbers of innovations (Acs & Audretsch 1989; Cohen and Levin 1989), since R&D expenditures capture the resources allocated to not only the innovation process but also the larger research production process.

In the same vein, patents provide a better measure of invention than innovation. For example, industries exhibiting a low propensity to patent, such as software, may exhibit a high level of innovative activity. Evidence indicates that the propensity to patent varies considerably across industries, that many innovations are never patented and that many patents have little economic value (Mansfield 1984; Scherer 1983; Shepherd 1979).[18] Consequently, patents as a measure of innovation may miss a not-insignificant segment of innovative activity. For example, over 34 percent of surveyed SBIR firms in a 1992 SBA report indicated that intellectual property protection was not needed for their newly innovated products (U.S. Small Business Administration 1992). It is, therefore, no wonder that debate continues on the appropriateness of patents as a measure of innovative output (Acs and Audtretsch 1989; Griliches 1990).

The most direct measures of innovation, actual innovation counts and citations to innovations, eliminate such drawbacks. Yet, compiling actual innovation counts is a time consuming process that few have attempted. A handful of industry-specific case studies offer insights into innovation in a narrow set of industries but do not provide a consistent, widespread source of data (Enos 1962; Gellman Research Associates 1976; Hamberg 1963; Jewkes 1969). On a much larger scale, the Small Business Administration carried out a one-time survey of innovations for the U.S. in 1982 to create the Innovation Data Base.[19] While this effort provides data across numerous industries, it offers data for a single year only, so no systematic collection of innovation counts has taken place in almost twenty years. This void of information across time severely restricts the application of this data to empirical analysis, particularly as time grows farther away from 1982; moreover, this data cannot shed any light on how innovation has changed over time.

Another issue concerning innovation measures relates to the level of data availability. Many innovation measures (such as R&D expenditures) are collected through government-sponsored surveys, and data cannot be disaggregated at small units of observation, such as the firm or geographic regions smaller than the state, due to data suppression. This inherently weakens the scope of these measures in empirical analysis.

In short, empirical examination of the innovation process has been hindered by data that imperfectly measure true innovation, lack observations across time, or cannot be geographically disaggregated. As noted earlier, Griliches (1990, p. 1669) sees the "dream of getting hold of an output indicator of inventive activity" as "one of the strong motivating forces for economic research in this area."

The SBIR Phase II award offers distinct advantages as a measure of innovative activity for small firms. First, a necessary condition for receipt of a Phase II award is that the firm has a feasible research project with the goal of commercialization that underwent review through the Phase I process. Phase II awards are similar to patents in this regard, in that they are an intermediate step towards a commercialized innovation. Yet, Phase II awards differ substantially from patents because, as a result of their strong relation to commercialization, they more closely approximate a final innovation.

The likelihood of a commercial outcome from SBIR research is substantial. The SBA (1992) reports that nearly 30 percent of Phase II projects early in the program achieved, or had plans to likely achieve, commercialization within four years of receiving the Phase II award. A 1991 survey by GAO of early Phase II projects found similar patterns of commercialization; 35 percent had achieved actual sales from SBIR research and almost 50 percent had received Phase III (non-SBIR) funding to continue development (U.S. General Accounting Office 1998). In 1996 DOD surveyed its SBIR award recipients from the 1990s and reported similar commercialization rates. Just under half of DOD's SBIR projects resulted in Phase III activity, with 32 percent already generating sales by 1996 (U.S. General Accounting Office 1998). This evidence suggests that at least one in three Phase II projects achieve commercial sales relatively soon, while approximately one in two projects are active in development.

Second, the Phase II award also offers a unique measure for examining the innovation mechanism of small, high-tech firms. The Program is mandated to target firms having 500 or fewer employees and solicits projects in high technology areas. SBIR firms are typically young and small. Over 41 percent of surveyed firms in the SBA 1992 report were less than five years old at the time of their Phase I award and nearly 70 percent had 30 or fewer employees. SBIR firms also concentrate most of their efforts on R&D. Over half the firms in the survey devoted at least 90 percent of their efforts to R&D.

Third, annual data on the SBIR Program are readily available for empirical analysis for 1983 onwards. The data are maintained at the research project level, providing information on the research topic, participating firm, funding agency, award amounts, geographic identifiers, and principal investigator (PI). This large sample spanning almost twenty years allows for both time series and longitudinal analysis since an individual firm's participation in the Program can be tracked across time. The type of data collected also opens up the possibility of examining innovation at the firm level as has traditionally been done, as well as at the project or individual (PI) level.

A significant strength of SBIR data is their ability to allow for the study of geographic patterns of innovation. The data contain the address of the firm awarded funding for each SBIR project. Hence, it is feasible to aggregate Phase II awards across a wide range of spatial areas, from zip codes to states. Unlike some innovation measures such as R&D expenditures, Phase II awards can be aggregated at the metropolitan or urban level. This is particularly relevant for examining the location of innovative activity, since spatial theory expounds the importance of urban areas in fostering knowledge and providing access to other productive resources (Lucas 1993).

The Phase II award also suffers from limitations as a measure of innovation, as do other measures of innovation. The most significant drawback to the use of SBIR data is the lack of industrial classification of firms participating in the SBIR Program. To gather this information requires extensive effort in matching SBIR data with other data sources. The relatively small number of SBIR awards as compared to other innovation measures, such as patents, may also restrict the application of this data to analyses requiring large numbers of observations. By the nature of the SBIR Program, the Phase II award serves exclusively as a measure of small-firm innovation and thus cannot measure innovation occurring in large firms.

Issues related to if and why firms pursue SBIR funding, and possible biases in the awards process invite caution and care in using SBIR data as well. It may be that there is underlying selectivity in determining which firms try to participate, and also succeed, in the SBIR Program. For instance, participation may be influenced by how well firms in a region are informed of the program's existence or by the prevailing tendency for firms in a region to engage in innovative behavior.

Moreover, the success of SBIR proposals depends to a great extent on the needs and interests of the funding agencies. Worthwhile projects may be overlooked if they are not part of a prevailing "hot" area of research that dominates an agency's research agenda.[20]

Recent efforts targeting the geographic distribution of SBIR activity may also weaken the Phase II awards as a measure of innovation; areas with traditionally low levels of innovation may be experiencing growing SBIR activity, dampening the correlation between Phase II awards and innovations.[21] Many of these limitations, however, can be overcome, making the SBIR Phase II award a useful measure of innovative activity among small firms in the United States.

SUMMARY

The Small Business Innovation Research Program emerged in the late 1970s out of a growing concern among policymakers that small business was being overlooked in federal R&D activities and that the global competitiveness of the U.S., particularly in high-tech sectors, was waning. Underlying these concerns was the strong belief that small business was a significant contributor to economic growth.

The Congress, therefore, enacted the SBIR Program in 1982 requiring every federal agency with a sizeable extramural R&D budget to set aside a portion of this budget to fund short-term, early-stage R&D at small firms. In addition to helping meet federal R&D needs, the intent was to stimulate innovation—and therefore economic activity—among high-tech small firms. The SBIR program is now the largest federal R&D initiative targeting small business.

It has been shown that the geographic concentration of SBIR activity closely resembles that of other measures of innovative activity. There is little difference in the regions of highest concentration between R&D expenditures, utility patents, and SBIR Phase II awards. Innovative activity tends to concentrate along the east and west coasts, with California and Northeastern states being dominant. This regional concentration is also seen at the metropolitan level. Within the most innovative states, certain metro areas account for most of the activity: San Francisco, Los Angeles, New York City, and Boston.

It has also been argued that the SBIR Phase II award can serve as a useful measure of innovative activity among high-tech small firms. While existing measures, such as R&D expenditures and patents, provide insight into innovative activity, their limitations call for the search for new measures. The strengths of the Phase II award as a measure lie in its focus on small firms, availability of data across geographic regions and time, and strong link to commercialized innovation. Therefore, the Phase II award can be used as an indicator of innovation to examine differences in small-firm innovative activity across industries and geographic regions.

3

GEOGRAPHY AND INNOVATION
The Role of Knowledge Spillovers and Agglomeration

The relationship between geography and innovative activity has been of longtime interest to economists. Traditional linkages have focused on agglomeration. More recently, as economists have become increasingly interested in the role of spillovers in economic growth, attention has also turned to the relationship between localized knowledge spillovers and innovative activity. But much remains to be learned. As Malecki (1983, p. 95) states, "Innovation may be the most important and the least understood aspect of the concept of spatially unbalanced growth."

This chapter examines the way agglomerative economies and knowledge spillovers influence localized innovative activity. The first section of this chapter explores agglomerative economies. The second section examines the role of localized knowledge spillovers in innovation and develops the knowledge production function. The final section of this chapter defines the local technological infrastructure as a system combining the sources of agglomeration and knowledge spillovers.

The discussion in this chapter argues that innovative activity is influenced by agglomeration and knowledge spillovers. What is important to realize is that these spillover effects can be geographically bounded so that their impact on innovation diminishes as distance increases from the source of these spillovers. It is the underlying sources of these spillovers that comprise the local technological infrastructure. Therefore, understanding the geographic concentration of innovative activity requires recognition and understanding of the local technological infrastructure.

AGGLOMERATION

Since Weber (1929), agglomeration economies have been held to be a major contributor to the clustering of firms in urban areas.[22] High technology firms, for instance, indicate that they choose locations with proximity to skilled labor, academic institutions, and favorable economic climates (Rees and Stafford 1986). Agglomeration economies arise when firms experience positive externalities associated with their proximity to institutions or other firms that affect their productivity. Agglomerative economies take the form of two related but distinct effects: localization economies and urbanization economies (Malecki 1991; O'Sullivan 2000). Localization economies revolve around industry-specific agglomeration, while urbanization economies are more general in nature, crossing industrial boundaries. With the combined effects of localization and urbanization economies, innovating firms have strong incentives to cluster together to take advantage of the various positive agglomeration economies spawned by geographic proximity (Bania et al. 1993). The geographic concentration of innovative activity is the consequence of the clustering of these innovative firms.

Localization economies occur largely from concentrations of labor and knowledge spillovers, both of which are particularly relevant to high-tech, small firms.[23] Firms can benefit from reduced innovation costs generated by lower labor costs if the search for and acquisition of labor is easier due to the proximity of a relevant labor pool (Angel 1991; Glasmeier 1986, 1988; Malecki and Bradbury 1992; Rosenthal and Strange 2001; Storper 1982). This suggests why many industries requiring certain types of skilled workers are clustered geographically. Dumais et al. (1997) find that manufacturing firms in the United States since 1970 have attempted to locate near similar firms using the same type of labor. This results in firms locating in areas with large labor pools; and, as more firms congregate in the same area, the pools grow larger.[24] The larger the available pools of labor relevant to particular industries, the greater the positive externality.

This labor agglomeration can be especially beneficial to high-tech firms requiring highly skilled and trained workers (Glaeser 2000; Glasmeier 1988; Rees and Stafford 1986; Satterthwaite 1992). Recent evidence indicates that the ability of firms to incorporate new technology depends on the availability of skilled labor, which tends to concentrate in large cities (Bartel and Lichtenberg 1987; Wozniak 1984, 1987). Therefore, the innovative activity in a region, such as a metropolitan area, may be greater with the presence of a sufficient and relevant labor pool. On the other hand, a concentration of labor that is not applicable to the innovative activities of firms in an area offers no agglomerative economies for these firms. For this reason, the strength of the link between innovation and agglomeration is not due to mere size but also to the source of the agglomeration. For firms to realize agglomerative economies due to available labor, the local labor pool must be relevant to the innovative activity of the firms.

Urbanization economies, on the other hand, exist because of positive externalities solely related to the size of a geographic area. Areas with large populations generate specialization of resources. This specialization breeds a broad business climate conducive to successful innovation (Beeson 1992; Jacobs 1960; Maleki 1983; Rees and Stafford 1986). Firms, for instance, can benefit from concentrated economic activity through the availability of local business services (MacPherson 1991; Markusen et al. 1986; Satterthwaite 1992; Saxenian 1985). Firms rely on services provided by other firms in the operations of their business, including legal counsel, printing, financing, or transportation. Higher concentrations of such services provide greater access for firms requiring external support in their operations. Coffey and Polese (1987) find that specialized service providers locate near clients. Therefore, the cost of innovation is lowered by the quickness and ease of acquiring services nearby instead of searching at farther distances, which increases the risk and length of the search process.

The size of an area can also provide agglomerative economies through greater access to networks among individuals, firms, and institutions located in the area. The opportunity for increased communication and cooperation among these agents enhances the flow of information useful to the innovative process and the ability to perform some types of innovative activity. Fritsch and Lukas (1999), for example, examine formal cooperation in three German regions and find that the tendency to cooperate is strongest between agents in the same region. They find that cooperation takes place through information trading, the sharing of resources (such as large-scale research labs) needed for innovation, contractual research arrangements, and formal joint ventures.

KNOWLEDGE SPILLOVERS

It is widely accepted that knowledge influences economic activity. Moreover, it has long been argued that economic agents rely not only on their own knowledge but also appropriate knowledge from other sources, whether it be codified or in tacit form. While the acquisition of both types of knowledge are influenced by geographic proximity, geography arguably plays a larger role in transmitting tacit knowledge because, by definition, tacit knowledge requires face-to-face interaction to be communicated. In addition to geographic proximity, technological proximity also plays a role in the transmission of knowledge. One reason for this is that industries sharing a similar technology are more likely to know the same code and to be aware of similar areas ripe for innovation. A more obvious reason is that industries related technologically likely share common incentives for innovation.

It follows that the firm's stock of knowledge depends in part on how close, in terms of geographic and technological space, the firm is to other firms and institutions with useful knowledge. Cities provide one avenue for closeness in geographic and technological space. According to Glaeser (2000, p. 83), cities are the "centers of idea creation and transmission." Lucas (1988) agrees, contending that cities are the breeding ground of knowledge spillovers. The clustering of knowledge sources stimulates an environment where knowledge can be more accessible and more easily transferred, whether formally or informally. Cities, Glaeser (2000) argues, facilitate the flow of ideas between individuals and firms, so that firms have an incentive to cluster geographically in order to learn from one another. This coincides with von Hippels' (1994) idea that knowledge is "sticky," in the sense that the cost of transmitting knowledge rises with distance. This implies that firms can reduce costs by locating near sources of knowledge and that knowledge spillovers are likely stronger in closer proximity to their source.

The mechanisms by which knowledge spills over received little attention until the late 1970s and 1980s. As the high-tech sector and knowledge-based industries began to rapidly expand, attention was drawn to understanding why knowledge matters and how it flows between economic agents. Wood (1969) first incorporated 'informational linkages' into the traditional agglomeration framework, while economists have separately explored the implications of external knowledge flows to the production process (Arrow 1962; Krugman 1991a, 1991b; Romer 1986, 1990; Schumpeter 1942).[25] Knowledge flows from both private and public sources, and much attention has been focused on measuring the extent of spillovers from these sources (Audretsch and Stephan 1996; Jaffe et al. 1993; Malecki 1985; Mansfield 1995; Rauch 1990). Jaffe et al. (1993), for example, find that patent citations are strongly linked to research performed in close proximity to the patenting firm. Surveying 66 firms across 7 industries and over 200 academic researchers, Mansfield (1995) finds that the percentage of firms located in the same state as universities they cite for contributions to their research is significantly related to the frequency of citation. He also finds the likelihood of firms to fund applied research at universities within 100 miles is far greater than that for universities at least 1,000 miles away, holding faculty quality constant.

In the private sector, spillovers emanate from information shared between firms, particularly those in related industries, which rely on a common pool of knowledge. Industrial R&D activity is a frequent source of knowledge transfer among firms. As the concentration of R&D activity increases in an area, the availability of knowledge increases as information about this R&D is disseminated. The link between R&D spillovers and innovative output has been well documented (Adams 1998; Griliches 1992; Hausman et al. 1984; Malecki 1985). Adams (1998) finds that 71 percent of firms surveyed in 1997 incurred expenditures to learn about related industrial R&D activities. Moreover, external sources of R&D increase these firms' patent applications and patents granted. Examining the concentration of innovation in U.S. metropolitan areas in 1982, Feldman and Audretsch (1999) find that firms in related industries sharing a common science base tend to cluster together geographically and to generate innovative activity; only 4 percent of new innovations in 1982 did not originate in metropolitan areas. Increased formal transfer of knowledge from R&D, sharing of R&D resources, and joint R&D activities are shown by Fritsch and

Lukas (1999) to be closely tied to increased innovative activity in three German regions.

Knowledge generated by research in the academic community is the predominant source of spillovers from the public arena for firms seeking knowledge. Universities perform approximately half of all federally funded basic research in the United States (National Academy of Sciences 1999). It is argued that knowledge flows relatively freely from public sources, and therefore, can be captured by innovating firms at lower costs than acquiring privately generated knowledge (Adams 2001). Consistent with this argument is the finding by Narin et al. (1997) that individuals associated with universities, government, or other public institutions authored 73 percent of research papers cited by U.S. patents. The majority of the U.S. cited papers originated from top research universities.

Universities have traditionally disseminated knowledge to the public through the mechanisms of publication, seminars, consulting, and education.[26] Universities also train graduates who provide easy access to cutting-edge knowledge, particularly technology of a tacit nature. Moreover, informal relationships with university researchers can be easier to form than with researchers at other firms. Many of these relationships benefit from close proximity. It is not surprising, then, that innovative activity is found to benefit from the geographic presence of a research university. The importance of proximity is not just that knowledge spills over to existing firms in an area. Firms also have the incentive to locate near research universities to acquire knowledge useful to their innovative activities. Moreover, universities often stimulate the birth of new industries and firms—examples of such spillovers abound. Dorfman (1983) and Saxenian (1985, 1996) indicate a long-lasting and strong link between MIT and the innovative activity in Boston, and Stanford and Silicon Valley. A recent study shows that biotechnology firms in Ohio and Sweden have clustered near research universities (Carlsson and Braunerhjelm 1999).

Evidence also points to a strong link between the positive effect on innovation of proximity to universities and the tendency of firms to cluster geographically. Bania et al. (1993) examine university spillovers in twenty-five metropolitan areas in the late 1970s and find significant evidence of spillovers in the evolving electronics industry. Mamuneas (1999) finds evidence of strong localized spillovers from publicly funded R&D in six high-tech industries in the U.S. As a measure of innovation, patent activity in high-tech industries is shown to increase due to local university spillovers (Jaffe 1989).

Knowledge Production Function

To empirically estimate the existence of knowledge spillovers and agglomeration economies in knowledge-based industries, past research (Acs et al. 1994; Anselin et al. 1997, 2000; Audretsch and Feldman 1996b; Feldman 1994a, 1994b; Jaffe 1986, 1989) has utilized what has come to be known as the 'knowledge production function.' Griliches (1979) first articulated the use of a production function to model the production of knowledge outputs as a function of knowledge inputs in an effort to estimate the returns to R&D. His knowledge production function included a measure of external knowledge available to firms in order to explicitly capture the spillover of knowledge between firms and industries. In developing this production function of knowledge, Griliches focused only on the influence of proximity in technological space, ignoring geographic space. In this framework, the effect of proximity is connected to the closeness of economic agents in terms of technological relatedness. For example, this means that a biotech firm would generally benefit more from knowledge generated by another biotech firm than by a chemical firm regardless of location.

Griliches outlined a Cobb-Douglas production function where the knowledge inputs are measures of R&D capital. He included a measure of knowledge spillovers in addition to the 'own' knowledge inputs to capture the influence of external knowledge capital on productivity. Equation 3.1 presents the knowledge production function constructed by Griliches:

$$Y_i = \beta X_i^{1-\gamma} K_i^{\gamma} K_a^{\upsilon}, \qquad (3.1)$$

where Y_i is a measure of the ith firm's innovative output, X_i is a vector of conventional inputs, K_i is own knowledge capital, and K_a is aggregated knowledge capital for the ith firm's industry. In this model, Griliches made three assumptions: (1) there are constant returns in the firm's own inputs; (2) aggregate knowledge capital is the sum of all firms' R&D capital in the given industry ($K_a = \sum K_i$); and (3) own resources are optimally allocated and all firms within an industry face the same relative input prices. The third assumption implies that the ratio between the knowledge and conventional inputs is:

$$\frac{K_i}{X_i} = \frac{\gamma}{1-\gamma} \frac{P_x}{P_k} = r, \qquad (3.2)$$

where P_x and P_k are prices of inputs X and K. Equation 3.2 indicates that the K/X ratio, r, is independent of an individual firm, i. This implies that K_i/X_i is constant and equal to r for all firms. It follows then that $\sum K_i / \sum X_i$ also equals r.

Aggregating Equation 3.1 for all firms within an industry and substituting $\sum K_i / \sum X_i$ into this equation yields the following:

$$\sum_i Y_i = \beta \left(\frac{\sum K_i}{\sum X_i} \right)^{\gamma} K_a^{\upsilon} \sum X_i. \qquad (3.3)$$

Given that $K_a = \sum K_i$, Equation 3.3 can be rewritten as:

$$\sum_i Y_i = \beta \left(\sum X_i \right)^{1-\gamma} K_a^{\upsilon+\gamma}. \tag{3.4}$$

Equation 3.4 shows the production function aggregated at the industry level, a useful model for analyzing productivity across industries.

It should be noted, as Griliches points out, that the coefficient for K_a (aggregated knowledge capital) is greater at the industry level than at the firm level because the aggregated level "reflects not only the private but also the social returns to research and development" (1979, p. 103). Mathematically, the coefficient $(\gamma+\upsilon)$ is greater than the coefficient γ, since γ and υ are positive. This larger effect of knowledge occurs because knowledge spills over and is thus available to an entire industry and not just an individual firm.

Griliches concedes that this model grows more complex when redefining the model to span multiple industries. The complication arises from the fact that firms (or industries) acquire knowledge in varying degrees from other firms (or industries), depending on their distance in technological space from the knowledge sources. Reinterpreting Equations 3.1 through 3.4 so that i indexes industries rather than firms, Griliches redefined aggregate knowledge capital as

$$K_{a_i} = \sum_j w_{ij} K_j. \tag{3.5}$$

Equation 3.5 states that the ith industry's aggregate knowledge (K_{ai}) is equated to a weighted sum of knowledge acquired from all available sources (K_j). Griliches interpreted the weight, w_{ij}, as the proportion of knowledge from source j borrowed by industry i. The weighted influence of this transfer of knowledge is assumed to grow stronger the closer in technological space an entity is to the source of the spillover. For instance, a firm should benefit more from the R&D efforts of other firms in the same four-digit SIC classification than from efforts by firms in the same two-digit classification.

Geographic Proximity

The knowledge production function put forth by Griliches has proven effective in the search for knowledge spillovers by providing a framework to model the relationship between knowledge inputs and knowledge-based outputs. Empirical research emanating from Griliches' model began in 1989 with Jaffe's pivotal look at the effect of academic R&D activity on corporate patenting. His premise (supported by his findings) was that knowledge generated within universities spills over and is economically useful for private-sector innovation. Jaffe's model of corporate patent activity included two inputs: industrial R&D and academic R&D.

Jaffe expanded upon Griliches' model to capture spillovers not only in technological space but also in geographic space. He estimated the strength of knowledge spillovers at the state level across selected industries. Jaffe recognized the state as a second-best unit of analysis, contending that geographic effects likely take place in a smaller area. Lack of disaggregated industrial R&D data, however, forced the unit of observation to the state level. He attempted to account for any error in the spatial unit being too broadly defined by including in the estimation a 'geographic coincidence index,' which tries to capture input concentrations within a state. His findings indicate that corporate patenting benefits from the R&D efforts of industries and universities in close proximity geographically and technologically, suggesting that knowledge indeed spills over not only from the private sector but also from the public sector.

Subsequent efforts have estimated the importance of geographic proximity to innovation at less aggregated levels in order to explore the complexity of the spillover mechanism across space. Recent studies (for example, Anselin et al. 1997, 2000; Feldman and Audretsch 1999) have focused on metropolitan areas. It is argued that it is within concentrated urban centers—not broad geographic regions— that agglomerative economies and knowledge spillovers most easily occur. The clustering of economic activity stimulates the development of new firms and attracts existing firms, thereby creating an atmosphere conducive to the growth of agglomerative economies and knowledge spillovers.

Empirical evidence suggests that close proximity to relevant knowledge leads to increased innovation in urban areas, though the results are far from definitive. According to Feldman and Audretsch (1999), innovative activity tends to cluster in metropolitan areas where complementary industries that share a common scientific knowledge base are concentrated. Anselin et al. (1997, 2000) find mixed results across industries in terms of the importance of geographic proximity to the number of new innovations when focusing on four high-tech industries in metro areas with innovative activity. The geographic effects associated with knowledge spillovers and agglomeration are more pronounced in the electronics and instruments industries than in chemicals and machinery. This lends credence to the belief that the spillover mechanism is indeed complex, varying across geographic space and industries.

THE LOCAL TECHNOLOGICAL INFRASTRUCTURE

The idea of the 'local technological infrastructure' lies in the efforts of researchers to capture the complete system of agglomerative and knowledge spillovers found in a geographic area. Carlsson and Stankiewicz (1991, p. 111) aptly define the technological infrastructure as "a particular infrastructure made up of academic institutions, research institutes, financial institutions, government agencies, and industry associations" in which "networks of agents (firms, organizations, and individuals) ... interact with each other." In other words, the elements comprising this infrastructure are the combined sources of agglomeration and knowledge inputs. Therefore, the technological infrastructure is the framework through which agglomeration and knowledge spillovers influence innovation.

Much research has focused on particular elements of this infrastructure (such as concentrations of labor or R&D), while far less has attempted to focus on the infrastructure itself. The literature that has explored the infrastructure as a whole largely describes the state of the infrastructure in innovative areas and makes implications based on this description about the relationship between the technological infrastructure and innovative activity (Dorfman 1983; Saxenian 1985, 1996; Fosler 1988; Smilor et al. 1988).

Interest in empirical analysis of the impact of the local technological infrastructure on innovation has heightened in recent years. However, the empirical model put forth by Jaffe (1989) is viewed as too restrictive to model the full effect of spillovers from the technological infrastructure on innovation. Building upon Jaffe's efforts, Feldman (1994a, 1994b) expands the range of inputs to also capture agglomeration effects, as well as knowledge spillovers, in the innovation process. In addition to industrial and academic R&D, Feldman includes measures of the concentration of informal networks created by entities in related industries and the concentration of business services that provide services relevant to successful commercialization. This expanded set of elements constitutes a measure of the local technological infrastructure. Due to data limitations, Feldman, like Jaffe, is restricted to state-level analysis. Unlike Jaffe, she uses actual innovation counts instead of patents as her output measure and focuses on the thirteen most innovative industries in her sample.[27] Feldman identifies a positive and significant spillover from academic and industrial R&D on innovation and finds evidence of positive externalities from agglomeration.

The benefits derived from spillovers may depend on firm size. Large firms have the capacity to internalize (at least some) knowledge production and scale economies reducing the need to appropriate knowledge from other sectors such as universities. Small firms, on the other hand, have limited internal resources, such as R&D labs, and may turn to external sources of knowledge. In support of this differential effect by firm size, Feldman (1994a, 1994b) notes significant differences between small and large firms in their reliance on agglomeration and spillover effects. Specifically, small firms benefit more from proximity to university research and business services, suggesting small firms rely more heavily on external sources of knowledge in innovation than large firms in the same industry.

SUMMARY

This chapter has reviewed the theoretical relationship between innovative activity and spillovers from knowledge and agglomeration. The local technological infrastructure encompasses the sources of agglomeration and knowledge spillovers. Positive externalities emanating from the local technological infrastructure can foster increased innovative activity within a geographic area. It has been established that geographic proximity influences the strength of these spillovers. Agglomeration of resources, particularly labor supply and business services, provides incentives for firms to cluster geographically and to innovate. Knowledge spills over from both private and public arenas. The most significant source of private knowledge is industrial R&D activity, while universities serve as the bastions of public knowledge. Proximity to the source of knowledge spillovers matters for innovative activity.

4

EVALUATING INNOVATIVE ACTIVITY

This chapter presents an empirical methodology for investigating the importance of knowledge spillovers and agglomerative economies to innovative activity. Specifically, a knowledge production function is defined for use in a hurdle model for count data. The knowledge production function provides a framework for estimating the impact of geographical and technological spillovers on innovative output. In particular, the model defines innovative output as a function of inputs emanating from key elements of the local technological infrastructure, including private industry and academe. The hurdle model allows for a two-step analysis of the impact of the technological infrastructure on first the likelihood of innovative activity and then the level of activity if it occurs. A negative binomial equation is employed in the second step of the hurdle model to account for the distributional characteristics of count data. To estimate the hurdle model, a unique data set is constructed based on SBIR Phase II awards as the measure of innovation and measures of the local technological infrastructure.

The first section of this chapter describes the empirical model to be estimated and the hypothesized impact of each component of the local technological infrastructure on innovative activity. A detailed description of the data to be used in the empirical analysis follows in the next section. The chapter concludes with a discussion of the econometric techniques employed to empirically estimate the knowledge production function using a hurdle model for count data.

EMPIRICAL MODEL OF THE LOCAL TECHNOLOGICAL INFRASTRUCTURE

This analysis follows Feldman (1994a, 1994b) in employing a knowledge production function to model the relationship between innovative activity and the local technological infrastructure.[28]   In this knowledge production function framework, a measure of knowledge output indicating some form of innovative activity is dependent upon a set of knowledge inputs, as well as sources of agglomerative economies. This general relationship is seen in Equation 4.1:

$$Y_{is} = KI_{is}^{\alpha} KA_{is}^{\beta} AG_{s}^{\delta} \qquad (4.1)$$

where Y is a measure of a knowledge output; KI is aggregated knowledge from the industrial sector; KA is aggregated knowledge from the academic sector; AG is aggregated sources of agglomeration inputs; α, β, and δ are parameter coefficients; i indexes industry; and s indexes geographic area.   The function expressed in Equation 4.1 states that innovative output within a given industry and geographic area is a function of knowledge originating from the private and academic (public) sectors in that industry and geographic area, as well as the effects of agglomeration.

Decomposing Equation 4.1 into specific knowledge and agglomeration components yields the general-form function shown in Equation 4.2:

$$INN_{is} = f(R\&DLABS_{is}, \ UNIV_{is}, \ EMPCON_{is}, \ EMPSIC73_{s}, \ POPDEN_{s}), \quad (4.2)$$

where INN is some measure of innovative activity; R&DLABS is industrial R&D activity; UNIV measures industry-related academic knowledge; EMPCON is the concentration of related industries; EMPSIC73 is the concentration of relevant business services; POPDEN is population density; i indexes industry; and s indexes the geographical unit of observation.   Substituting these variables into Equation 4.1, the knowledge production function to be estimated is:

$$INN_{is} = R\&DLABS_{is}^{\beta_1} UNIV_{is}^{\beta_2} EMPCON_{is}^{\beta_3} EMPSIC73_{s}^{\beta_4} POPDEN_{s}^{\beta_5}. \quad (4.3)$$

Table 7 defines the variables used in the empirical estimation.   It should be noted that in the empirical estimation private sector R&D cannot be aggregated at the industry level due to the inability to disaggregate the R&D labs data by industry. Therefore, R&DLABS is only disaggregated at the geographic area level.

As proposed in Chapter 2, the Phase II SBIR award is employed as the primary measure of innovative output. Two variables are constructed to indicate innovative activity: PH2DUM and PH2. Consistent with the hurdle model, whether or not a geographic area has experienced any innovative activity related to a given industry is captured by PH2DUM. PH2, on the other hand, indicates the rate of innovation in areas that experience at least some level of innovative activity.

Table 7. Definition of Variables

POPDEN$_s$	Average number of persons per square kilometer in a metro area, 1990-95
R&DLABS$_s$	Average number of R&D labs located within a metro area, 1990-95
EMPSIC73$_s$	Average employment level in business services (SIC 73) within a metro area, 1990-95
EMPCON$_{is}$	Average location quotient for employment in industry i within a metro area, 1990-95
UNIVDUM$_s$	Dummy variable indicating whether (=1) or not (=0) any Research I/II or Doctorate I/II universities were located in a metro area, 1990-95
UNIVR&D$_{is}$	Total level of academic R&D expenditures by Research I/II or Doctorate I/II institutions in fields corresponding to industry i within a metro area, 1990-95 (in thousands of 1992 Dollars)
PH2DUM$_{is}$	Dummy variable indicating whether (=1) or not (=0) a metro area had at least one firm receive any Phase II SBIR awards in industry i, 1990-95
PH2$_{is}$	Total number of Phase II SBIR awards received by firms in industry i within a metro area, 1990-95

Private sector R&D activity is reflected by the number of R&D labs within a metropolitan area. Industrial R&D expenditures, commonly used as a measure of R&D activity, are unavailable at the metropolitan level, and therefore, are not used in this study. Instead, R&DLABS is used as a proxy for knowledge generated by industrial R&D. It is argued that as industrial R&D activity increases, firms have a larger pool of knowledge to draw upon in the process of creating innovations. The ability to access this knowledge is positively related to geographic proximity. Therefore, innovative activity is expected to increase as more R&D labs appear in an area.

Two variables are constructed to measure knowledge emanating from the academic sector: UNIVDUM and UNIVR&D. The first, UNIVDUM, is a zero-one dummy variable and provides an indication of access to knowledge from the academic community. UNIVR&D takes this one step further by proxying the amount of relevant knowledge produced at universities as measured by academic R&D activity. Academic R&D activity is restricted to R&D expenditures by universities classified as Carnegie Research I/II or Doctorate I/II institutions. This subset of universities is selected because it is these institutions that are responsible for the bulk of research in the U.S., thereby making them the predominant academic source of knowledge within a region (National Science Board 2000). Knowledge contributed by other academic institutions likely plays a much smaller role in the knowledge spillover process. Moreover, Research I/II and Doctorate I/II institutions, through graduate programs, generate the highly trained science and engineering workforce, a major source of tacit knowledge for firms hiring their graduates. Given the high correlation between R&D expenditures and conferred degrees in science and engineering fields, these institutions' R&D expenditures in science and engineering fields proxy the knowledge embodied in human capital as well as in research.

The presence and concentration of related industry is captured in a location quotient that defines an industry's employment concentration in a metropolitan area relative to its national concentration. The location quotient is defined as follows:

$$LQ_i = \frac{E_i / \sum_i E}{N_i / \sum_i N} * 100 , \qquad (4.4)$$

where E is employment within a metropolitan area in industry i and N is national employment in industry i. Benchmarked at 100, a location quotient greater than 100 indicates a metro area that has a relatively high concentration of employment in an industry compared to the United States overall, while a value less than 100 indicates a concentration of employment in that industry below the national average. The higher the concentration of employment, the greater the potential knowledge transfers between firms in the same industry.

Employment in SIC 73, business services, measures the concentration of relevant business services. Business services include a broad array of services offered to firms, including advertisement, printing services, computer programming, data processing, personnel services, and patent brokerage. Higher levels of employment indicate increased availability of services used in the innovation process. Therefore, a high employment level in business services in a metro area should benefit local innovative firms, contributing to higher innovative activity.

Population density is included in the model to isolate the influence of the size of an area on innovative activity. The more populated an area is, the more likely are individuals engaged in innovative activity are to encounter other individuals who hold useful knowledge and appropriate that knowledge through personal relationships. Increased population density, therefore, is expected to positively affect innovative activity within a metropolitan area. In a more general sense, population density is also an indication of the size of local economic activity. As a metropolitan area becomes denser, it is expected that overall economic activity will increase due to the rising necessity and desire for economic interaction among individuals and institutions in the area. Population density, therefore, also captures the positive effect of agglomerative economies on innovation in areas with substantial levels of economic activity.

The unit of observation is narrowed to metropolitan areas in the United States to focus on agglomeration and spillover effects from the local technological infrastructure. The sample includes all 273 distinct metropolitan areas located in the United States.[29]

To control for differences in the effects of the local technological infrastructure on innovation across industries, five industries are examined: Chemicals and Allied Products (SIC 28), Industrial Machinery (SIC35), Electronics (SIC 36), Instruments (SIC 38), and Research Services (SIC 87). The industries are grouped at the two-digit SIC code level. These industries were selected because they encompass the vast majority of high-technology fields where most innovative activity occurs. Appendix A lists the four-digit industry classifications comprising these five two-digit industries. Chemicals includes pharmaceutical and other biological products. Industrial machinery covers firms producing a wide range of machinery and computer hardware. Firms producing most electronic equipment and electrical components, including semi-conductors, fall within the electronics industry. Instruments includes any type of controlling and analyzing device or instrument, from navigational equipment to process controls, to medical equipment. Firms primarily engaged in contractual R&D, engineering, and management-consulting services are in Research Services.

DATA

The number of SBIR Phase II awards at the metropolitan level, $PH2_{is}$, was aggregated from firm level data available from the Small Business Administration, which maintains an annual data set of awards from the Program's inception in 1983. A significant drawback to this data set is its lack of industrial classification for participating firms. To classify these firms by industry required considerable effort. Every firm was first searched for in the CorpTech Database, and if identified, the 4-digit SIC code was recorded. A sizeable number of firms were not listed in the CorpTech Database, however, these unidentified firms were then investigated in two steps: first using *Ward's Business Directory*, 1990-95, and next using individual Internet searches for each firm. Every effort was made to reduce the possibility of misidentification, including verifying a firm's address or location. Firms identified using the Internet were assigned an SIC code at the two-digit, three-digit, or four-digit level depending on reported information.[30] Industry affiliations or SIC codes were explicitly listed for some firms; however, for many firms deduction was required to determine industrial classification. SIC codes were assigned based on reported business descriptions, main products, and research agendas. This extensive, three-step effort resulted in over 90 percent of recipient firms being classified in an industry.

The National Science Foundation's WebCASPAR provided institutional level data on academic R&D expenditures by department for Carnegie Research I/II and Doctorate I/II institutions. Institutions were linked to a metropolitan area and academic R&D expenditures were determined by assigning departments to the relevant industry. Following Feldman (1994b), academic R&D expenditures were determined by linking each industry to relevant science and engineering departments based on academic field classifications from the National Science Foundation's Survey of Research and Development Expenditures at Universities and Colleges. Table 8 shows the links between academic departments and industries used to determine the relevance of academic R&D expenditures to industries.

Table 8. Links Between Academic Departments and Industries

Industry	Academic Department
Chemicals and Allied Products (SIC 28)	Medicine, Biology, Chemistry, and Chemical Engineering
Industrial and Commercial Machinery and Computer Equipment (SIC 35)	Electrical Engineering, Astronomy, Physics, Computer Science, Mechanical Engineering and Other Engineering and Physical Sciences
Electronic and Other Electrical Equipment and Components (SIC 36)	Electrical Engineering, Astronomy, Physics, and Computer Science
Measuring, Analyzing and Controlling Instruments (SIC 38)	Medicine, Biology, Electrical Engineering, Astronomy, Physics, Computer Science, Mechanical Engineering and Other Engineering and Physical Sciences
Research Services (SIC 87)	All science and engineering fields

Source: Feldman, Maryann. *The Geography of Innovation.* Dordrecht, The Netherlands: Kluwer Academic Publishers, 1994.

Counts of R&D labs were collected from the annual *Directory of American Research and Technology*.[31] This data required considerable effort to aggregate to the metropolitan level. The directory lists R&D labs by city and state. R&D labs were counted for each city by state. Each city was then matched to a metropolitan area, and the counts by city were summed for each metro area. The average number of R&D labs was approximately 37, while counts ranged from 0 to over 1,300 across the 273 metro areas.

Industrial employment data were compiled from the Bureau of the Census' *County Business Patterns*. County level data were then aggregated to the metropolitan level. Employment in business services, EMPSIC73, ranged from a low of 211 to over 402,000, with a mean near 20,000. In machinery and electronics, the average employment concentration (EMPCON$_i$) was near the benchmark of 100. The mean was noticeably smaller than 100 in instruments and research services. Only in chemistry was the average employment concentration well above 100.

Population statistics for metropolitan areas were taken from the Bureau of Economic Analyses' *Regional Economic Information Systems, 1969-96*. Population density is defined as the ratio of population to the land size (in kilometers) of a metropolitan area.

Table 9 reports descriptive statistics for the sample that includes all 273 metropolitan areas in the United States for the aggregated period, 1990-95. Aggregation allows for reliable statistical analysis across industry groups. The figures in Appendix B show the geographic distribution of each variable across metropolitan areas. The distribution of Phase II awards across metro areas is highly skewed towards low counts; only a handful of metropolitan areas receive large numbers of awards. During 1990-95, the average number of Phase II awards received within a metro area ranged from 1 in machinery to 6.5 in research services. The means across the five industries are quite small due to the high proportion of metro areas with no Phase II awards. For example, one-third of U.S. states receive approximately 85 percent of all SBIR awards, leaving only 15 percent of awards to be received by the majority of states (Tibbetts 1998).

Tables 10 and 11 report descriptive statistics for the variables comprising the technological infrastructure across two categories of metropolitan areas: those with and without SBIR Phase II activity in 1990-95. (Appendix C lists the number of Phase II awards for each metropolitan area.) What is evident from these tables is that Phase II activity is split between U.S. metropolitan areas, with 136 metro areas receiving no Phase II awards during 1990-95 and 137 receiving awards, indicating that half of the metropolitan areas in the United States generated little innovative output using Phase II awards as a proxy of innovative activity.

Second, there is considerable difference in the technological infrastructures of these two groups of areas. Table 12 summarizes and compares the means across the two groups. The means for all but the employment concentrations in chemicals, industrial machinery, and electronics are significantly different between the two groups.[32] The average metropolitan area with no Phase II awards is relatively small, has disproportionately low concentrations of employment in the scientific instruments and research services industries, has almost no R&D labs and has low levels of academic R&D expenditures. Moreover, areas without Phase II activity are highly likely not to have even any presence of research universities, in stark contrast to metropolitan areas with Phase II activity.

Table 9. Descriptive Statistics for All Metropolitan Areas, 1990-95 (N=273)

Variable	Mean	Std. Dev.	Minimum	Maximum
POPDEN	101.2	103.9	1.9	954.0
R&DLABS	36.8	118.7	0.0	1,322.8
EMPSIC73	19,120.4	48,588.9	211.1	402,672.7
$EMPCON_{28}$	121.1	218.9	0.0	1,555.8
$EMPCON_{35}$	104.0	102.9	1.3	694.3
$EMPCON_{36}$	104.5	149.2	0.0	1,573.0
$EMPCON_{38}$	82.0	134.4	0.0	1,049.9
$EMPCON_{87}$	79.5	54.9	0.0	508.9
UNIVDUM	0.3	0.5	0.0	1.0
$UNIVR\&D_{28}$	132,420.1	462,262.6	0.0	4,062,545.1
$UNIVR\&D_{35}$	61,406.5	230,820.6	0.0	2,347,983.4
$UNIVR\&D_{36}$	37,883.6	161,325.5	0.0	1,799,800.1
$UNIVR\&D_{38}$	176,196.6	624,531.6	0.0	5,025,893.3
$UNIVR\&D_{87}$	234,651.6	778,714.4	0.0	6,163,377.4
$PH2_{28}$	1.4	5.9	0.0	51.0
$PH2_{35}$	1.0	4.4	0.0	46.0
$PH2_{36}$	2.7	12.4	0.0	132.0
$PH2_{38}$	3.5	14.5	0.0	138.0
$PH2_{87}$	6.5	30.3	0.0	371.0

Table 10. Descriptive Statistics for Metropolitan Areas Receiving No Phase II
Awards, 1990-95 (N=136)

Variable	Mean	Std. Dev.	Minimum	Maximum
POPDEN	72.4	86.0	1.9	954.0
R&DLABS	4.8	6.3	0	31
EMPSIC73	3,878.4	4,661.2	211.1	32,001.2
$EMPCON_{28}$	140.1	258.2	0	1,555.8
$EMPCON_{35}$	105.3	102.9	5.1	526.9
$EMPCON_{36}$	91.3	154.0	0	1,573.0
$EMPCON_{38}$	64.6	130.7	0	785.7
$EMPCON_{87}$	56.3	26.3	0	173.9
UNIVDUM	0.09	0.3	0	1.0
$UNIVR\&D_{28}$	3,556.6	15,191.8	0	93,945.2
$UNIVR\&D_{35}$	3,104.6	18,261.6	0	174,485.3
$UNIVR\&D_{36}$	1,633.0	9,580.8	0	92,496.7
$UNIVR\&D_{38}$	5,401.2	28,384.1	0	232,237.2
$UNIVR\&D_{87}$	11,072.8	52527.7	0	406,943.4
$PH2_{28}$	0	0	0	0
$PH2_{35}$	0	0	0	0
$PH2_{36}$	0	0	0	0
$PH2_{38}$	0	0	0	0
$PH2_{87}$	0	0	0	0

Table11. Descriptive Statistics for Metropolitan Areas
Receiving Phase II Awards, 1990-95 (N=137)

Variable	Mean	Std. Dev.	Minimum	Maximum
POPDEN	129.7	112.3	9.8	744.5
R&DLABS	68.6	161.6	0	1322.8
EMPSIC73	34,251.2	65,094.3	418.8	402,672.7
$EMPCON_{28}$	102.2	170.0	0.7	1399.9
$EMPCON_{35}$	102.8	103.1	1.3	694.3
$EMPCON_{36}$	117.5	143.5	0.8	953.6
$EMPCON_{38}$	99.3	136.3	0	1049.9
$EMPCON_{87}$	102.5	65.4	28.3	508.9
UNIVDUM	0.51	0.5	0	1.0
$UNIVR\&D_{28}$	260,342.9	627,737.2	0	4,062,545.1
$UNIVR\&D_{35}$	119,282.8	315,359.9	0	2,347,983.4
$UNIVR\&D_{36}$	73,869.6	222,130.7	0	1,799,800.1
$UNIVR\&D_{38}$	345,745.3	849,205.8	0	5,025,893.3
$UNIVR\&D_{87}$	456,598.5	1,053,775.7	0	6,163,377.4
$PH2_{28}$	2.8	8.1	0	51.0
$PH2_{35}$	2.1	6.0	0	46.0
$PH2_{36}$	5.3	17.2	0	132.0
$PH2_{38}$	7.0	19.9	0	138.0
$PH2_{87}$	13.0	41.9	0	371.0

Table 12. Difference in the Means for Metropolitan Areas
by SBIR Phase II Activity, 1990-95

Variable	Metro Areas With No Phase II Awards (N=136) Mean	Metro Areas With Phase II Awards (N=137) Mean
POP**	235,969.6	1,258,695.3
PDEN**	72.4	129.7
R&DLABS**	4.8	68.6
EMPSIC73**	3,878.4	34,251.2
$EMPCON_{28}$	140.1	102.2
$EMPCON_{35}$	105.3	102.8
$EMPCON_{36}$	91.3	117.5
$EMPCON_{38}$*	64.6	99.3
$EMPCON_{87}$**	56.3	102.5
UNIVDUM**	0.09	0.51
$UNIVR\&D_{28}$**	3,556.6	260,342.9
$UNIVR\&D_{35}$**	3,104.6	119,282.8
$UNIVR\&D_{36}$**	1,633.0	73,869.6
$UNIVR\&D_{38}$**	5,401.2	345,745.3
$UNIVR\&D_{87}$**	11,072.8	456,598.5
$PH2_{28}$**	0	2.8
$PH2_{35}$**	0	2.1
$PH2_{36}$**	0	5.3
$PH2_{38}$**	0	7.0
$PH2_{87}$**	0	13.0

*Difference in the means is significant at the five percent level.
**Difference in the means is significant at the one percent level.

The average metropolitan area receiving Phase II awards looks strikingly different. This metropolitan area is large, with a population over 1.2 million, and is almost twice as densely populated as the average metro area with no Phase II activity. Industrial employment concentrations are relatively on par with the United States as a whole (seen by the average location quotients being near 100). Academic R&D activity is substantially greater than in the typical metropolitan area with no Phase II activity. Academic R&D expenditures range from 38 to 73 times greater in the average metropolitan area with Phase II awards than in those without Phase II awards. The enormous differences in most components of the technological infrastructure between areas with and without Phase II activity offer preliminary evidence in support of the hypothesis that the technological infrastructure affects local innovative activity.

ECONOMETRIC ESTIMATION

A hurdle model for count data (Mullahy 1986; Pohlmeier and Ulrich 1992) is used to carry out the empirical estimation in this study. The hurdle model allows for systematic differences in the statistical processes leading to zero and positive observations, which may exist in a research-funding program such as the SBIR Program. In other words, the hurdle model allows for the separate examination of the impact of observable characteristics on the cluster of observations with zero activity and the frequency of activity for observations with some level of positive activity. In this case, the hurdle model provides a means to investigate the potentially different effects of knowledge spillovers and agglomerative economies on whether or not innovation occurs, as well as the rate of innovation in areas experiencing innovative activity. The model, therefore, distinguishes between the factors that influence the 'participation' decision from those that influence the 'frequency' decision in the innovation process.

The hurdle model allows for this separation of effects by using a two-step estimation procedure. The first step estimates the likelihood of whether or not one or more Phase II awards are received by firms in a metropolitan area using a binary choice model (here, a probit). This is done by estimating the probability that at least one firm in a metropolitan area receives at least one Phase II award dependent on the local technological infrastructure. The probit model takes the form:

$$z_i^* = \beta' X_i + \varepsilon, \text{ where } \varepsilon_i \sim N[0,1], \tag{4.5}$$

$$z_i = 1 \text{ if } z_i^* > 0 \text{ and } z_i = 0 \text{ if } z_i^* \leq 0,$$

where z_i^* is an unobservable latent variable related to the presence and SBIR activity of small, high-tech firms, and z_i is a binary choice variable indicating whether or not any Phase II awards are received, which is observable. The probit equation estimates the production function defined in Equation 4.3 using PH2DUM as the dependent variable. The dependent variable (z_i=PH2DUM$_i$), therefore, is a dummy variable indicating whether (=1) or not (=0) a metropolitan area has at least one firm that receives any Phase II SBIR awards. The probit equation also measures academic R&D activity by UNIVDUM, a dummy variable signifying whether (=1) or not (=0) a metro area has any research oriented universities to capture the importance of simply having such institutions, regardless of the *level* of their research activity in a given field.

The second step of the hurdle model estimates the effect of observable characteristics on the frequency decision, here the rate of innovation. Because Phase II awards are a non-negative, integer measure of innovative activity, a technique accounting for the distributional characteristics of count data is employed.[33] A negative binomial model instead of the common Poisson model is estimated because overdispersion is apparent in the data. The appropriateness of the negative binomial model is verified by a test to determine if overdispersion exists. This test is performed by estimating a Poisson model and using the estimated coefficients in the calculation of α, the overdispersion parameter. The overdispersion parameter is calculated as follows:

$$\alpha = \frac{(y_i - \lambda_i)^2 - \lambda_i}{\lambda_i \sqrt{2}}, \tag{4.6}$$

where λ_i is the estimated value of $\exp[X\rho B]$ (Greene 1993). A t-test is performed to determine if α is significantly different from zero. The null hypothesis is that $\text{var}(y_i)=E(y_i)$; the alternative hypothesis that $\text{var}(y_i) =E(y_i)+ \alpha g(E(y_i))$. Rejection of the null hypothesis is consistent with overdispersion and the application of the negative binomial model; non-rejection of the null supports the use of the Poisson model. In this study, the estimates of α are significantly greater than zero, indicating the negative binomial model is better suited than the Poisson for these data.

Following Cameron and Trivedi (1998), the negative binomial equation in the second step of the hurdle model takes the form:

$$f(y_i \mid \lambda, \alpha) = \frac{\Gamma(y+\alpha^{-1})}{\Gamma(y+1)\Gamma(\alpha^{-1})}\left(\frac{\alpha^{-1}}{\alpha^{-1}+\lambda}\right)^{\alpha^{-1}}\left(\frac{\lambda}{\alpha^{-1}+\lambda}\right)^{y}, \quad y_i = 0, 1, 2, \ldots \text{ and } \alpha \geq 0, \tag{4.7}$$

where λ is the mean given by

$$E(y_i \mid x_i) = \lambda_i = \exp(x_i' \beta). \tag{4.8}$$

The negative binomial distribution relaxes the Poisson condition that the mean equals the variance so that the variance is given by

$$\omega_i = V(y_i \mid x_i) = \omega(\lambda_i, \alpha) = \lambda_i + \alpha\lambda_i^2, \tag{4.9}$$

where α is a scalar parameter. In the case of overdispersion, as is evident in this analysis, the mean (λ) is less than the variance (ω).

Unobserved factors related to small firms in a metropolitan area may generate sample selection in the SBIR analysis. Unobservable characteristics of a metropolitan area could affect the existence and innovative activity (including SBIR activity) of small high-tech firms in the area. Areas with similar technological infrastructures could experience differences in the development of small business sectors and resulting innovative activity given variation in local entrepreneurial climates (Malecki 1991; Martinez and Nueño 1988). Selection could also arise if unobserved factors related to markets for research funds or the SBIR Program influence innovative activity differently across metropolitan areas. For instance, capital markets that are often geographically bounded may influence a firm's knowledge about, and choice of, alternative types of capital—including the SBIR Program. Moreover, the prevalence of information about the SBIR Program varies across metropolitan areas, so that firms in areas with similar technological infrastructures may have drastically different levels of knowledge about the SBIR Program. In addition, unobservable characteristics of the structure of the SBIR Program could also induce sample selection. Thus, knowledge about the SBIR Program, its structure, and the local entrepreneurial climate could affect the SBIR participation of small firms at the metropolitan area level.

If sample selection exists and is ignored, estimation will yield biased parameter estimates (Manksi 1995). Because of potential selection, an approach to correct for sample selection in the Heckman (1979) tradition is incorporated into the model following Greene (1994). This two-step estimation process naturally fits into the hurdle model framework. The Inverse Mills Ratio is estimated from the results of the probit equation and then included as an explanatory variable in the negative binomial equation to estimate the significance of selection in the model. The standard errors in the second-step, negative binomial equation are corrected using the method outlined by Murphy and Topel (1985). Correction for selectivity alters the interpretation of the second-step equation in the hurdle model. In the uncorrected model, this equation is conditional on metropolitan areas with positive Phase II counts (see Chapter 4). With the inclusion of the Inverse Mills Ratio for selectivity correction, the equation becomes unconditional, so that interpretation of estimated coefficients is based on all metropolitan areas regardless of the level of Phase II activity and not only on those with positive Phase II counts.

Identification occurs because the independent variables in the first- and second-step equations are not identically defined. While both equations model technological infrastructure, the variables related to university research differ in the two equations to capture different effects of university research. The probit equation includes a dummy variable, $UNIVDUM_s$, to indicate the presence of research universities in a metropolitan area, and the negative binomial equation includes the level of academic R&D expenditures in industry-related fields, $UNIVR\&D_{is}$. The probit variable addresses the importance of the existence of research universities, and the negative binomial variable, the intensity of research from these universities if present.

SUMMARY

This chapter has outlined the methodology used in this study to estimate the effects of knowledge spillovers and agglomerative economies on innovative activity. Following past research, a knowledge production function is employed to examine the effects of knowledge spillovers and agglomerative economies on the innovative activity of high-tech small firms at the metropolitan area level. A unique data set is constructed using Phase II awards from the SBIR Program as the primary measure of innovative activity, and a hurdle model for count data is used to econometrically estimate the spillover and agglomeration effects related to the innovation process. The hurdle model uses a two-step procedure, so that the effect of the local technological infrastructure is first estimated on whether or not innovative activity occurs (probit equation) and then on the rate of innovation (negative binomial equation).

The empirical findings are reported in Chapters 5 and 6. Chapter 5 presents the findings for Phase II activity by industry. Chapter 6 explores whether or not the spillover process is related to the federal agency providing the SBIR funding for Phase II. Chapter 7 presents empirical results for patent activity, where patents serve as the measure of innovation. Chapter 7 broadens the scope of this analysis by examining the local technological infrastructure's effect on the patent measure of innovation and thus provides the means for comparison between Phase II activity and patenting.

5

METROPOLITAN SBIR ACTIVITY IN THE 1990s

The previous chapter outlined the empirical methodology employed in this study to estimate the importance of spillovers from the local technological infrastructure on innovative activity. A knowledge production function was defined to capture key elements of the technological infrastructure, including concentrations of industrial activity and business services, academic R&D activity, industrial R&D activity, and area size. Restating Equation 4.3, this knowledge production function is defined as:

$$INN_{is} = f(R\&DLABS_{is}, UNIV_{is}, EMPCON_{is}, EMPSIC73_s, POPDEN_s), \qquad (5.1)$$

where INN is a measure of innovative output; R&DLABS is industrial R&D activity; UNIV is academic research activity; EMPCON is the concentration of employment in related industries; EMPSIC73 is the concentration of employment in business services; POPDEN is population density; i indexes industry; and s indexes the geographical unit of observation. In short, this knowledge production function defines innovative activity as a function of key sources of knowledge spillovers and agglomerative economies found in the local technological infrastructure.

This chapter presents findings for the empirical estimation of the effects of the local technological infrastructure on innovative activity. The hurdle model outlined in Chapter 4 is employed to estimate the impact of these effects first on whether or not innovative activity takes place in a metropolitan area, and second, on the level of innovation. This chapter focuses on SBIR Phase II awards as the measure of innovative output to capture innovation within the high-tech small business sector. Chapter 7 will present empirical results where utility patents in the private sector are the measure of innovative activity to examine whether the local technological infrastructure has a differential impact on innovative activity using a different measure and a different population of innovating agents.

As described in Chapter 3, previous empirical studies indicate that a link exists between localized knowledge spillovers and innovative activity. Yet, only a handful of studies have attempted to disaggregate this link by firm size (Acs and Audretsch 1993, 1996; Acs et al. 1994; Feldman 1994b), and few others have focused on the effect of these spillovers at the metropolitan area level (Anselin et al. 1997, 2000; Feldman and Audretsch 1999). Moreover, these studies have focused on the state of innovation in 1982, relying almost exclusively on the 1982 Innovation Data Base described in Chapter 2. The measures of innovation employed in this study, Phase II awards and patents, provide a means of comparison to past research that used these 1982 innovation counts data.

The empirical findings reported in this chapter indicate that knowledge spillovers and agglomerative economies play a significant role in the innovative process of small firms and that this role varies across industries. Local knowledge spillovers and agglomerative economies within a metropolitan area have a more significant effect on the likelihood of receiving Phase II awards than on the number of Phase II awards received. Proximity to R&D labs and research universities has the most consistent, positive effect on the likelihood of receiving Phase II awards across industries. The concentrations of business services and industrial employment also significantly impact the probability of receiving Phase II awards, though these effects are industry specific.

There is less evidence of the importance of localized spillovers in determining the number of Phase II awards (or the rate of innovation) for small firms at the metropolitan area level. While the role of the academic community remains strong, with the level of university R&D activity having a positive and significant effect on the number of awards, the role of R&D labs and business services vanishes in determining the level of Phase II activity. More local R&D labs or a higher concentration of business services has no significant effect on the number of Phase II awards. Moreover, the concentration of industry employment has a significantly positive effect in only one of the industries.

THE LIKELIHOOD OF PHASE II ACTIVITY

Table 13 presents the estimated results from the first-step, probit equations of the hurdle model for five high-tech industries. The results indicate that knowledge spillovers, and to a lesser extent agglomerative economies, positively affect whether or not small firms within a metropolitan area are the recipients of one or more SBIR Phase II awards. The presence of research universities (UNIVDUM) is positive and highly significant across all five industries.[34] Thus, areas where firms can appropriate knowledge from the academic community into their innovation process

Table 13. Probit Equation of the Likelihood of Phase II Awards

Variable	Estimated Coefficients (Standard Errors)				
	Chemicals & Allied Products	Industrial Machinery	Electronics	Instruments	Research Services
Constant	-1.8174 (0.2017)	-1.6282 (0.2227)	-1.7269 (0.1763)	-1.7559 (0.1915)	-1.6467 (0.2229)
POPDEN	0.2229** (0.1031)	-0.1985 (0.1823)	-0.02726 (0.1236)	-0.008945 (0.1482)	-0.2032 (0.1753)
R&DLABS	2.1743*** (0.5718)	0.9722** (0.5118)	1.9766*** (0.5531)	1.9755** (0.8226)	2.2901*** (0.6758)
EMPSIC73	-0.1566** (0.07819)	0.1441* (0.09459)	0.05241 (0.09401)	0.2360** (0.1323)	0.01730 (0.09562)
UNIVDUM$_i$	1.02763*** (0.2325)	0.5787** (0.2477)	0.5994*** (0.2402)	0.7900*** (0.2350)	0.9464*** (0.2264)
EMPCON$_i$	-0.03651 (0.07231)	0.05087 (0.1139)	0.08114 (0.06648)	0.1831*** (0.07460)	0.7770*** (0.1924)
Ln L	-80.4	-83.0	-85.1	-89.1	-106.2
χ^2	86.9	113.7	112.1	140.9	136.5

*Significant at the 10 percent level
**Significant at the 5 percent level
***Significant at the 1 percent level

are more likely to see some level of Phase II activity than areas where firms do not have access to nearby research universities, indicating that proximity to research universities is an important component in the decision to innovate for high-tech small firms. The number of R&D labs (R&DLABS) is also significantly related to the probability of a metro area having firms that receive Phase II awards. Metropolitan areas with more labs have a higher probability of Phase II activity occurring, suggesting that small firms within these areas benefit from proximity to clusters of industrial R&D activity.

The findings for UNIVDUM and R&DLABS are consistent with previous evidence of small firms' reliance on external knowledge flows in the innovation process. Acs et al. (1994) and Feldman (1994b) find that both academic and industrial R&D have a significant effect on the number of small firm innovations and that the effect of universities on small firms' innovative activity is greater than that for large firms. Anselin et al. (1997, 2000) find similar evidence for high-tech firms at the metropolitan area level in four of the same industries examined here. These findings are also supported by evidence on the role of geographic proximity between patents and patent citations. Jaffe, Trajtenberg, and Henderson (1993), for example, find that citations to other patents are significantly more likely to refer to patents with inventors in the same metropolitan area.

The findings for the significance of agglomerative economies are less consistent than for knowledge spillovers. Proximity to related industry has mixed effects in its impact on the receipt of Phase II awards in a metro area. The concentration of employment (EMPCON) in chemicals, machinery, and electronics has no significant impact on the likelihood of firms within a metropolitan area receiving one or more Phase II awards. For instruments and research services, however, a higher concentration of industry-specific employment leads to a significant increase in the probability of receiving Phase II awards. A strong proximity effect in research services is expected given that these services are often collaborative and provided to third parties. This can lead to frequent interaction between firms in this industry, allowing small firms to more easily appropriate knowledge from other firms doing similar work.

The impact of the prevalence of business services within a metropolitan area (EMPSIC73) on the likelihood of receiving Phase II awards also varies across industries. In electronics and research services, there is no significant impact of increased employment in business services on the receipt of Phase II awards. The lack of a relationship in research services may be expected given the nature of the industry; these firms provide contractual R&D services to other firms and may not seek the same level of business services as firms engaged in innovative activity predominately for themselves. A strong, positive effect exists in instruments, and a less significant but positive impact is found in machinery, suggesting small firms in these industries more heavily rely on external services in the innovative process. An unexpected negative and significant relationship exists in chemicals between business services employment and the likelihood of receiving Phase II awards. In this case, firms engaged in SBIR activity (predominately biotechnology) may require either less business services altogether or only certain types of services in the innovative process compared to SBIR firms in other industries.

The density of the metropolitan area (POPDEN) is not significantly related to the likelihood of Phase II activity in four of the five industries, indicating the probability of innovation is not driven by area size.[35] Innovation in terms of Phase II awards, then, does not occur in large urban areas simply because of the size of the area. This suggests agglomerative economies arising from the scale and scope of the local economy play little role in the likelihood of Phase II activity among small firms at the metropolitan area level. This is not the case in the chemical industry. An increase in the number of people per square kilometer leads to a significant increase in the probability of a metropolitan area having Phase II activity in the chemicals industry. This result may be driven by peculiarities in the biotech industry where networks of individuals and inter-firm interaction play a significant role (Audretsch and Stephan 1996; Henderson and Cockburn 1996; Walcott 1999)

THE RATE OF PHASE II ACTIVITY

Table 14 shows the empirical results from the negative binomial equations in the second step of the hurdle model for the five high-tech industries. The overdispersion parameter, α, is significant for each industry, indicating that overdispersion is present in the data and that the negative binomial specification is more appropriate than Poisson for these data. The Inverse Mills Ratio is negative across all five industries and highly significant in four of the five industries, indicating that sample selection contributes to differences across metropolitan areas and must be taken into account to obtain unbiased parameter estimates.[36] For these four industries, the likelihood of receiving Phase II awards in a metropolitan area is non-randomly associated with the number of Phase II awards. In other words, unobservable factors influencing the receipt of any Phase II awards in a metropolitan area are negatively correlated with the unobservable factors affecting the number of awards received in that area. In the machinery industry, however, the Inverse Mills Ratio is insignificant, indicating selection bias is not present in this industry sample. This suggests there is no non-random correlation in machinery between the unobservable factors affecting the likelihood of Phase II activity and those influencing the number of awards.

The negative direction of the selection bias likely results from the net effect of several simultaneous unobserved factors, making it difficult to identify the sources of selection. Yet, it is useful to consider potential sources for the sake of future research that may be able to control for some of the currently unobservable factors. One reason for the negative selection could be that metropolitan areas with strong technological infrastructures and entrepreneurial climates are funded less well relative to other areas receiving SBIR funding. This would suggest a funding bias in the SBIR Program so that the 'have-nots' are more likely to be funded than the 'haves.' In addition, areas with freer and larger capital markets could experience fewer awards due to easier access to alternative sources of capital, even though these areas may have a higher likelihood of participation because of greater availability of information about the SBIR Program. The negative selection bias may also be indicative of the unmeasured effect of the number of participating firms in a metropolitan area. Metropolitan areas with fewer firms applying for SBIR funding

Table 14. Negative Binomial Equation for the Number of Phase II Awards

Variable	Chemicals & Allied Products	Industrial Machinery	Estimated Coefficients (Standard Errors) Electronics	Instruments	Research Services
Constant	1.9966 (0.3028)	1.1123 (0.5222)	2.3700 (0.4933)	2.08541 (0.3343)	1.5322 (0.3086)
POPDEN	-0.2816* (0.1550)	0.3716* (0.2415)	-0.1068 (0.1436)	-0.1446* (0.08844)	-0.01929 (0.1250)
R&DLABS	0.2111 (0.1654)	0.01118 (0.2811)	0.2908 (0.2391)	0.1965 (0.1684)	0.08835 (0.1889)
EMPSIC73	0.01066 (0.03150)	-0.01349 (0.05721)	-0.03806 (0.05195)	0.01147 (0.04832)	0.008336 (0.05574)
UNIVR&D$_i$	0.04108* (0.02498)	0.06474 (0.05228)	0.1369** (0.06549)	0.03013* (0.02068)	0.03962*** (0.01423)
EMPCON$_i$	-0.4179*** (0.1628)	-0.2322 (0.2501)	-0.05283 (0.1386)	0.02696 (0.06346)	0.6129*** (0.1423)
INVERSE MILLS RATIO	-0.4771*** (0.1740)	-0.1307 (0.2286)	-0.7098*** (0.2915)	-0.6071*** (0.2281)	-0.8522*** (0.1822)
α	0.5411*** (0.2150)	0.3018** (0.1454)	0.6368*** (0.2022)	0.5194*** (0.1638)	0.4933*** (0.1241)
Ln L	-117.4	-133.8	-172.1	-219.2	-279.5

*Significant at the 10 percent level
**Significant at the 5 percent level
***Significant at the 1 percent level

could yield fewer total awards across firms than areas with higher firm participation after controlling for the strength of the technological infrastructure.

What stands out across industries in Table 14 is the overall weaker contribution of the local technological infrastructure to the *number* of Phase II awards compared to the *likelihood* of receiving one or more awards seen in Table 13. Few of the independent variables are significant, and when they are, it is at a lower level. These results indicate that knowledge spillovers and agglomerative economies generate different effects during the innovation process. Spillovers more strongly influence whether innovative activity occurs and less the rate of innovation for high-tech small firms. The technological infrastructure within a metropolitan area is a key determinant of whether innovation occurs in that area but less of a factor in how much innovation takes place. This may account for past difficulty in documenting a strong link between knowledge spillovers and innovation when looking only at the rate of innovative activity.

The significance of industrial R&D activity diminishes between the probit and negative binomial analyses of Phase II awards. The significant, positive effect of industrial R&D activity (R&DLABS) on the likelihood of Phase II awards disappears at traditional levels of significance when analysis shifts to the number of Phase II awards. This suggests industrial R&D is not vital to the rate of Phase II activity so that SBIR firms may rely on non-industrial sources for general knowledge appropriated into the innovative process. Considerable prior evidence indicates that small firms appropriate knowledge generated by local universities and that this knowledge is a key determinant of these firms' innovative activity. This holds for Phase II activity as well. Compared to the other components of the technological infrastructure, the frequency of innovation among small, high-tech firms depends most strongly on the magnitude of industry-related R&D activity performed by research-oriented universities (UNIVR&D). The positive and significant spillovers emanating from local universities persist in every industry but machinery. In electronics, UNIVR&D is the only significant variable, suggesting that proximity to related university R&D is vital to innovation in electronics.[37]

The significance of research universities, though, is weaker in the negative binomial equations compared to the probit equations for every industry except research services. Research universities play a greater role in determining the likelihood of innovative activity than the rate of innovation. For Phase II activity, these results suggest that proximity to research universities matters more in terms of providing access to knowledge than in the volume of knowledge provided, thereby allowing innovative activity to take place in areas where it otherwise would not. This coincides with the argument that local universities provide knowledge through mechanisms besides research, such as the education of the local workforce.

The concentration of industrial employment (EMPCON) is significant only in chemicals and research services, implying that the positive spillovers associated with proximity to similar firms matter less in determining the frequency of innovation than merely the presence of innovative activity. The concentration of employment in research services is positively related to the number of Phase II awards as expected. The reasoning is the same as that described for the positive effect of EMPCON on the likelihood of Phase II activity. Employment concentration in chemicals has an unexpected negative impact on the number of Phase II awards.

This implies for the chemicals industry that Phase II activity favors metropolitan areas with relatively low concentrations of employment suggesting large-scale clustering inhibits Phase II activity. In other words, proximity to 'too many' similar firms may lead to agglomerative diseconomies in the chemicals industry.

This negative impact may reflect industry scale effects driven by large chemical and pharmaceutical manufacturers that dominate the industry. These large firms dictate the high concentrations of employment in certain metropolitan areas, while the clustering of small firms in the industry may not be driven by proximity to these large firms. The SBIR firms in this two-digit industry group predominately fall in the biotechnology arena, which can be concentrated in quite different areas than those where large chemical manufacturers are located. Therefore, the negative effect estimated in this sample is at least in part the result of differences in the geographic distribution of firms in the chemical industry and the broadness of the industrial grouping at the two-digit SIC level.

Employment in business services (EMPSIC73) has no significant impact on the number of Phase II awards that firms within a metropolitan area receive. Proximity to available services has no discernable effect on the rate of innovative activity as measured by Phase II awards. This lack of a relationship holds across all five high-tech industries and may stem from the early-stage nature of SBIR research. Firms engaged in Phase II projects may not be at the stage of the innovation process that requires the use of external business services. Business services are often used later in the innovation process as development moves towards intellectual property protection and commercialization. For instance, firms close to the commercialization stage may seek legal and marketing services to successfully bring their product into the marketplace.

Size of the metropolitan area is a more decisive factor of the number of Phase II awards than the likelihood of Phase II activity, though the direction of its impact varies. Contrary to the *a priori* expectation, the negative effect of population density (POPDEN) on the number of Phase II awards is significant in chemicals and instruments; it is positively and significantly related in machinery. The insignificant effect in electronics and research services seen in the first-step probit analysis holds in the negative binomial estimation. If agglomerative economies exist, they do not spill over in a meaningful way to influence the number of Phase II awards received by small firms in electronics and research services.

The negative effect on the number of Phase II awards in chemicals and instruments suggests a negative agglomeration effect outweighs any positive externalities experienced by these industries. This result may not be as surprising as thought at first glance. One explanation is that rising costs due to increased competition for resources needed to successfully innovate may dampen the frequency of innovation for small firms in more densely populated areas if these costs are sufficient enough to reduce the incentives to engage in (more) innovative activity.

Interestingly, the direction of the effect changes in the chemical industry between the probit and negative binomial estimations. As a metropolitan area grows more dense, the likelihood of firms engaging in Phase II activity increases but the number of Phase II awards received by firms in the area declines. This suggests that small firms in the chemical industry benefit from more densely populated areas, perhaps because of the ability to expand networks and more easily gain information, and will be more likely to participate in the SBIR Program. It also suggests that areas with chemical-related firms engaged in Phase II activity experience sizeable agglomerative diseconomies, inclining these firms to reduce their Phase II activity because of increasing costs.

SUMMARY

This chapter has presented empirical estimates of the impact of the local technological infrastructure on innovative activity. Using the SBIR Phase II award as the measure of innovation, this impact was examined at two levels: (1) on the likelihood of innovative activity occurring in a metropolitan area and (2) on the level of innovative activity.

The evidence indicates that geographic proximity to the sources of knowledge spillovers and agglomeration plays a significant role in the innovative process and that this role varies across industries. This supports previous research with similar findings at both the state and metropolitan area levels for patents and innovation counts. Industrial R&D and research universities provide significant spillovers to high-tech firms, particularly in terms of their likelihood to engage in innovative activity. Agglomerative economies due to the clustering of industrial employment, business services, and general economic activity also influence innovative activity. The following chapter examines the impact of the government agency that funds SBIR activity on the effect of the local technological infrastructure.

6

AGENCY EFFECTS IN FEDERALLY FUNDED INNOVATION

The empirical results of Chapter 5 suggest that SBIR Phase II award activity depends on the local technological infrastructure and that the strength of this dependence varies across industries. The relationship between the technological infrastructure and SBIR activity may also be influenced by the structure of the SBIR Program itself. For example, Phase II activity may vary across the government agencies that fund the awards, reflecting the fact that certain agencies rely on the local technological infrastructure differently than other agencies. The two largest awarders of SBIR funding, DOD and HHS (largely NIH), generally focus on drastically different types of research. Moreover, these two agencies have historically targeted different types of researchers: DOD has longtime relationships with industry while HHS has strong connections with universities. These different "spheres of connectedness" may influence SBIR activity in at least two ways. First, each agency shapes its purpose and use of the SBIR Program based on its overall research plan and operations, and these can be dramatically different across agencies. Moreover, agencies may focus on different types of research, which in turn influence their selection of projects for SBIR funding. DOD, for example, tends to support product-oriented research in its overall research agenda. HHS, however, places substantial importance on knowledge-oriented research. These differences likely influence the approach these agencies take in their support of SBIR research. Second, the type of researcher that seeks SBIR funding may differ across agencies. For instance, many researchers seeking DOD-SBIR funding may come from or have backgrounds at defense-related companies, while many of those seeking HHS-SBIR funding may be in transition from academic research labs.

This chapter examines the impact of agency-specific effects on SBIR Phase II activity across metropolitan areas. A probit model capturing the impact of the local technological infrastructure on whether or not Phase II activity occurs is estimated for the top three SBIR funding agencies: the Department of Defense, the Department of Health and Human Services, and the National Aeronautical and Space Administration. These three agencies represent particular types of agencies and research agendas. DOD and NASA focus largely on industries related to materials, electronics, chemicals, and machinery, while HHS research centers on the biological and chemical industries. The probit model here is identical in structure to that employed in Chapter 5. The model is estimated separately for each of the three agencies by industry.

A drawback of disaggregating by agency is that the thinness of the data leads to non-convergence in the iterative estimation process in many instances, and hence, the negative binomial equation cannot be estimated for all agencies for each of the five industries.[38] Because of the lack of sufficient results for comparison across agencies, empirical results from the negative binomial estimations of the frequency of Phase II activity for the instances where convergence occurred are not reported in this chapter.[39] Instead, only the results from the probit estimations of the likelihood of Phase II activity are presented.

GEOGRAPHIC CONCENTRATION OF PHASE II ACTIVITY BY FUNDING AGENCY

SBIR Phase II activity varies across funding agencies, in terms of both the number of awards and the distribution of Phase II activity across metropolitan areas. Table 15 shows the number of Phase II awards funded by DOD, HHS, NASA, and all other agencies combined for the five high-tech industries. The top three agencies—DOD, HHS, and NASA—account for between 69 to 81 percent of all awards across the five industries, with DOD funding at least 45 percent of awards in all but chemicals and HHS funding the majority of awards in chemicals. Awards funded by NASA exceed those funded by HHS only in industrial machinery and electronics.

Table 15 also lists the number of metropolitan areas having Phase II activity by funding agency. It is clear that agency-level Phase II activity is concentrated across metropolitan areas. Among the top three agencies, Phase II activity occurred in one-fourth to one-third of all metropolitan areas. In the case of DOD, 93 of the 273 metropolitan areas (34 percent) received one or more Phase II awards, while HHS or NASA activity was found in approximately 70 metro areas (26 percent). At the industry level, Phase II activity funded by DOD was also noticeably more dispersed across metropolitan areas than that funded by HHS and NASA, with the exception of chemicals where HHS-funded activity was the most diffused. Phase II activity funded by the top three agencies was most concentrated in chemicals and industrial machinery. HHS-funded activity in industrial machinery, for instance, took place in only seven metropolitan areas; in chemicals, NASA-funded activity was found in only eight metro areas. This was most likely due to the overall low level of activity by these agencies in these industries.

Table 15. Number of SBIR Phase II Awards and Metropolitan Areas
with Phase II Awards by Funding Agency and Industry, 1990-95

	Number of Awards (Number of Metropolitan Areas with Phase II Activity)			
	DOD	HHS	NASA	Other Agencies
Chemicals &	72	213	15	80
Allied Products	(23)	(39)	(8)	(25)
Industrial	129	21	47	89
Machinery	(27)	(7)	(21)	(28)
Electronics	420	36	133	95
	(48)	(17)	(30)	(25)
Instruments	385	251	141	178
	(54)	(44)	(33)	(38)
Research	884	308	256	337
Services	(66)	(48)	(44)	(51)
High-Tech	1890	829	592	779
Industries	(93)	(71)	(70)	(72)

THE LIKELIHOOD OF PHASE II ACTIVITY

Tables 16-20 report the estimated results from the probit equations by agency for each of the five high-tech industries across all 273 metropolitan areas. Consistent with the results in Chapter 5, these findings show that the number of research labs and the presence of research universities in a metropolitan area generally have the most consistent significant impact on the likelihood of Phase II activity across industries. Agglomeration effects due to population density and the availability of nearby business services have no significant effect in four of the five industries.

Table 16. Probit Equation for Phase II Awards by SBIR Funding Agency
SIC 28 - Chemicals and Allied Products

	Estimated Coefficients (Standard Errors)		
Variable	DOD	HHS	NASA
Constant	-2.09535 (0.2968)	-1.8272 (0.2070)	-2.03342 (0.4377)
POPDEN	0.01119 (0.1725)	0.2272** (0.1015)	-0.8979* (0.4888)
R&DLABS	1.7618*** (0.5758)	0.3607 (0.3254)	1.09763* (0.6652)
EMPSIC73	-0.1535* (0.08968)	0.01891 (0.06056)	-0.06598 (0.1117)
UNIVDUM$_i$	0.6827** (0.3176)	0.7874*** (0.2395	0.8203* (0.4837)
EMPCON$_i$	-0.1512 (0.1935)	-0.08096 (0.09544)	-0.04912 (0.1726)
Ln L	-42.9	-78.7	-19.0
χ^2	72.0	66.5	34.3

*Significant at the 10 percent level
**Significant at the 5 percent level
***Significant at the 1 percent level

Of particular interest is the finding that within industries the results vary by agency. For example, in the chemical industry, only the presence of research universities consistently affects Phase II activity across all three agencies (see Table 16). Research labs have a positive and significant effect on DOD and NASA Phase II activity but an insignificant effect on HHS Phase II activity. The availability of business services has a significant and negative effect on Phase II activity for DOD but no significant effect for HHS and NASA. The size of the metropolitan area measured by population density has a differential impact on Phase II activity across all three agencies: significant and positive for HHS, significant and negative for

NASA, and insignificant for DOD. Surprisingly, the concentration of employment that occurs in the chemicals industry has no significant effect on the likelihood of Phase II activity, regardless of the agency.

Table 17. Probit Equation for Phase II Awards by SBIR Funding Agency
SIC 35 - Industrial Machinery

Variable	Estimated Coefficients (Standard Errors)		
	DOD	HHS[1]	NASA
Constant	-1.7966		-1.9759
	(0.2378)		(0.2723)
POPDEN	-0.005482		0.1089
	(0.1539)		(0.1268)
R&DLABS	0.4834		0.3573
	(0.3327)		(0.2905)
EMPSIC73	0.02049		0.02037
	(0.06620)		(0.06036)
UNIVDUM$_i$	0.3349		0.3729
	(0.2755)		(0.2936)
EMPCON$_i$	0.04798		-0.05609
	(0.1202)		(0.1646)
Ln L	-65.4		-53.0
χ^2	45.4		42.1

[1]Model inestimable due to perfect correlation between UNIVDUM and PH2DUM
*Significant at the 10 percent level
**Significant at the 5 percent level
***Significant at the 1 percent level

As seen in Table 17, there appears to be no significant impact of the local technological infrastructure on the likelihood of Phase II activity in the machinery industry for the two agencies for which results are reported. No HHS probit equation was estimated in this case due to perfect collinearity between UNIVDUM and PH2DUM in machinery.40 The probit results suggests that for Phase II activity funded by DOD or NASA, neither local knowledge spillovers nor agglomeration effects play a role in the likelihood of Phase II activity in the machinery industry. The small number of Phase II awards in machinery, however, may drive this result. Knowledge and agglomeration spillovers are apparent in the aggregated equations of Chapter 5 where R&D labs, availability of business services, and the presence of research universities significantly affect the likelihood of Phase II activity in machinery. The result may also be caused by the dominance of agencies other than DOD, HHS, or NASA in machinery-related Phase II activity. The limited Phase II activity of these three agencies in the machinery industry may cloud the role of the technological infrastructure.

Table 18 reports the estimated results for DOD, HHS, and NASA in the electronics industry. Stark differences appear between HHS and the other two agencies. The local technological infrastructure has no significant effect on the likelihood of HHS Phase II activity; the low level of HHS funding in electronics-related SBIR research may contribute to this apparent lack of effect. DOD and NASA exhibit similar patterns of influence between the local technological infrastructure and the likelihood of Phase II activity. Metropolitan areas with more R&D labs and a presence of research universities are more likely to see Phase II activity occur.

The strong positive effect of the technological infrastructure on overall Phase II activity in the instruments industry (seen in Chapter 5) diminishes considerably when Phase II activity is disaggregated by agency (see Table 19). The likelihood of HHS and NASA Phase II activity react in a significant manner only to the presence of research universities in a metropolitan area. No other source of spillovers has a significant impact on Phase II activity funded by these agencies. In contrast, the number of R&D labs, as well as the presence of research universities, positively and significantly impacts the likelihood of increasing DOD Phase II activity in a metropolitan area.

The agency results for the research services industry closely resemble the results in Chapter 5 and show little variation across agencies (see Table 20). This is likely due to the broad scope of SBIR research associated with this industry compared to the other more focused industries. The presence of research universities has the most significant impact on the likelihood of Phase II activity for all three agencies. Unlike the other four industries, the effect of the concentration of employment in research services is also significant and positive for DOD, HHS, and NASA. The only difference across agencies is the effect of R&D labs on the likelihood of Phase II activity. There is no apparent private R&D effect on the likelihood of HHS Phase II activity, but a significant and positive effect for DOD and NASA. This suggests that metropolitan areas with more research labs are more likely than other areas to experience greater DOD and NASA Phase II activity but negligible change in the likelihood of HHS Phase II activity.

Table 18. Probit Equation for Phase II Awards by SBIR Funding Agency
SIC 36 - Electronics and Electrical Equipment

| | Estimated Coefficients (Standard Errors) | | |
Variable	DOD	HHS	NASA
Constant	-1.7065	-1.9732	-1.8227
	(0.1898)	(0.2530)	(0.2285)
POPDEN	0.01131	-0.1455	0.02308
	(0.1341)	(0.1917)	(0.1678)
R&DLABS	0.8673**	0.5434	0.7195*
	(0.4613)	(0.3377)	(0.3781)
EMPSIC73	0.04413	-0.01022	0.02839
	(0.07475)	(0.06619)	(0.06436)
UNIVDUM$_i$	0.8502***	0.4329	0.5185*
	(0.2341)	(0.3053)	(.02782)
EMPCON$_i$	-0.001796	0.07532	-0.08035
	(0.07786)	(0.08728)	(0.1197)
Ln L	-84.5	-46.8	-63.2
χ^2	84.9	33.7	62.7

*Significant at the 10 percent level
**Significant at the 5 percent level
***Significant at the 1 percent level

Table 19. Probit Equation for Phase II Awards by SBIR Funding Agency
SIC 38 – Scientific Instruments

Variable	Estimated Coefficients (Standard Errors)		
	DOD	HHS	NASA
Constant	-1.5147	-1.9180	-1.8851
	(0.1855)	(0.2217)	(0.20813)
POPDEN	-0.1976	-0.1288	0.2638
	(0.1677)	(0.1684)	(0.1368)
R&DLABS	1.2003**	0.7405	0.4302
	(0.5180)	(0.4739)	(0.3265)
EMPSIC73	0.03959	0.08249	0.03132
	(0.08068)	(0.07801)	(0.06412)
UNIVDUM$_i$	0.8288***	1.1584***	0.6141**
	(0.2302)	(0.2535)	(0.2541)
EMPCON$_i$	0.07733	0.06726	0.1007
	(0.06555)	(0.07590)	(0.07070)
Ln L	-90.5	-70.3	-70.9
χ^2	90.5	100.6	59.4

*Significant at the 10 percent level
**Significant at the 5 percent level
***Significant at the 1 percent level

Table 20. *Probit Equation for Phase II Awards by SBIR Funding Agency*
SIC 87 - Research Services

Variable	Estimated Coefficients (Standard Errors)		
	DOD	HHS	NASA
Constant	-2.04114	-2.4100	-2.6396
	(0.2343)	(0.2853)	(0.3019)
POPDEN	0.002919	0.02983	0.1254
	(0.1302)	(0.1404)	(0.1234)
R&DLABS	0.6537*	0.6409	0.9203**
	(0.3749)	(0.4535)	(0.4515)
EMPSIC73	0.01899	0.07532	-0.07568
	(0.06625)	(0.07411)	(0.07146)
UNIVDUM$_i$	0.5085**	0.8300***	0.9582***
	(0.2231)	(0.2488)	(0.2523)
EMPCON$_i$	1.02778***	0.7458***	0.9299***
	(0.1932)	(0.2037)	(0.2088)
Ln L	-102.8	-73.4	-71.3
χ^2	96.4	107.1	98.5

*Significant at the 10 percent level
**Significant at the 5 percent level
***Significant at the 1 percent level

Past research (Finch 1987; Mehay and Solnick 1990; Markusen 1986; Weston and Gummett 1987) has explored the link between defense-related activities, the clustering of high-tech firms, and regional economic growth. Appendix D presents results of the probit model estimated using specifications that include a measure of the presence of nearby military installations. Two specifications were estimated: one including the number of military installations within the metropolitan area, and the other including the number of military installations in the state or states in which the metropolitan area is located. These specifications test whether or not agency-level Phase II activity is influenced by proximity to military operations. DOD-funded innovative activity may cluster close to military installations due to the more secretive and product-oriented nature of its research (Markusen et al. 1986; Malecki 1991; Mazza and Wilkinson 1980). While the defense industry is dominated by a few large firms, subcontracting by specialized small firms is prevalent and is geographically concentrated around the prime-contract firms (Gansler 1980; Rees 1981, 1982). Therefore, small firms engaged in DOD-sponsored research may find it necessary to have more direct contact with military personnel, so that proximity to military installations may be advantageous for the innovation process (Markusen 1986).

Tables 40 through 45 of Appendix D report empirical findings on the role of proximity to military installations in Phase II activity by agency for five industries and for the aggregated high-technology sector. It is strikingly evident that the number of military installations at the state level has no significant effect on the likelihood of Phase II activity regardless of the funding agency. This suggests that if proximity to a military presence matters, then proximity matters at the local level and not at the state level. This is confirmed by the results for the specifications including the number of military installations at the metropolitan level. The findings indicate a differential impact of military presence across agencies and industries. For the five industries combined as the high-tech sector, military presence within the metropolitan area has no significant effect on the likelihood of Phase II activity across all three agencies. Across all five individual industries, the number of local military installations has no impact on Phase II activity funded by HHS. For NASA, its effect is only significant for Phase II activity in the research services industry. However, the number of local military installations has a significant and positive effect on the likelihood of DOD-funded Phase II activity in three of the five industries—industrial machinery, instruments, and research services. There is no noticeable effect of proximity to local military installations in chemicals or electronics, even for DOD Phase II activity. These results suggest that proximity to military installations matters, but as expected, largely for DOD-related innovation. Moreover, the effect of proximity is confined to a local level and not larger regions, such as the state, and the importance of proximity to military installations depends on the industry related to the innovation.

SUMMARY

The results presented in this chapter indicate that variation in the relationship between the local technological infrastructure and Phase II activity is partly related to the funding agency. HHS' pattern of Phase II activity across industries differs considerably from the other agencies. The local technological infrastructure has a less widespread effect on the likelihood of HHS Phase II activity than for DOD or NASA. The impact of knowledge and agglomeration spillovers from the technological infrastructure on the likelihood of DOD and NASA Phase II are similar. On the other hand, the presence of research universities is consistently more related to SBIR activity for HHS than for the other two agencies. Moreover, a higher concentration of military installations is shown to be important for defense-oriented Phase II activity in certain industries. These findings, in conjunction with the insignificant effect of R&D labs on HHS Phase II activity, are consistent with the observation that HHS has a different sphere of connectedness than do DOD and NASA. The importance of research universities is consistent with both the knowledge-oriented nature of HHS projects as well as the apparent proclivity of HHS-SBIR recipients to have had a close connection with a university at one time. The DOD and NASA results, on the other hand, are consistent with the product-oriented nature of these agencies' research and the close connection these agencies have with industry.

These differences in the effects of knowledge spillovers and agglomeration across agencies suggest a note of caution to policymakers seeking to develop local innovative activity. The implication of these findings is that a single, broad-sweeping policy may not be most effective in stimulating SBIR activity. Policymakers should evaluate their local economy and determine what type of research and innovative activity is most suited to the structure of their economy. From this, policymakers can target SBIR funding agencies that focus on these lines of activity. Therefore, policymakers should develop economic policies that attract specific types of SBIR funding. For instance, defense- and space-related SBIR activity would benefit more from policies promoting the build-up of local R&D labs and a strong local military presence (particularly for defense-oriented activity). Health-related SBIR activity, on the other hand, would be better served by policies strengthening the local academic sector.

The results of this chapter also suggest the impact of the technological infrastructure grows less apparent as the unit of observation is more narrowly defined. In this case, the effect diminishes as SBIR awards by agency are further disaggregated by industry. The implication is that Phase II activity at the industry level likely benefits from spillovers more than Phase II activity disaggregated by agency in the same industry. In a broader context, this suggests that the benefits of spillovers are likely stronger for innovative activity at broader units of observations than smaller ones. This could also mean that spillovers related to innovative activity are more apparent at the industry level than at the firm level. This differential impact of spillovers may arise due to the composition of the local technological infrastructure, which influences the nature of resulting spillovers. The available pools of knowledge, for instance, are generally broad in scope, so that their relevance is more applicable to a range of activities than just a particular set of activities. Hence, a local pool of knowledge is likely more applicable to a range of related activities within an entire industry than to the specific activities of a single firm. The force of this implication, however, must not be overstated based only on the empirical results shown in this chapter. The less compelling evidence for significant spillovers and agglomerative economies at the agency level may be an artifact of the relatively low levels of Phase II activity disaggregated by agency across metropolitan areas that mask a more significant importance of spillovers and agglomerative economies.

7

METROPOLITAN PATENT ACTIVITY IN THE 1990s

The results for Phase II awards provide evidence that geographic proximity at the metropolitan area level plays a significant role in the innovation process of small firms. Whether or not a metropolitan area receives one or more Phase II awards strongly depends on knowledge spillovers from the presence of research universities and industrial R&D labs. The effects of geographic proximity to knowledge sources diminish in explaining the number of Phase II awards. The positive effect of industrial R&D labs disappears altogether, and the impact of academic R&D activity, as measured by size of expenditures, becomes weaker. Agglomeration economies (measured by population density, business services employment, and the concentration of industrial employment) matter to a lesser extent, having no consistent effect on SBIR activity across industries either in terms of the existence or degree of activity.

It is not clear, however, whether these effects at the metropolitan area level hold for other types of innovative activity or for firms of any size. These agglomeration and spillover effects may be unique to SBIR activity and may play a larger role for small firms, as Feldman (1994b) suggests. This chapter presents empirical findings for patent activity in the private sector to test whether geographic proximity at the metropolitan area level plays an important role in determining patent activity in high-technology industries. To the best of our knowledge, this analysis is the first to examine patent activity at the metropolitan area level by industry. It addresses whether differential effects of the local technological infrastructure exist across different types of innovative activity, particularly patents and SBIR Phase II awards. This study, however, cannot directly examine whether differential effects occur by firm size because the SBIR data pertain only to small firms and firm size is not known in the patent data.

The first section of this chapter describes the patent data set used for this empirical analysis. The next section discusses the geographic concentration of patent activity in high-tech sectors. The third section of this chapter reports the empirical findings on the impact of the local technological infrastructure on patenting activity in the United States in 1990-95, examining its effect first on the likelihood of patent activity occurring and then on the rate of patenting in metropolitan areas with patent activity. A comparison between these results and those for SBIR activity reported in Chapter 5 follows. The chapter concludes with a summary of the empirical findings and implications related to the impact of the local technological infrastructure on patent activity.

PATENT DATA

The empirical analysis in this chapter estimates the knowledge production function outlined in Equation 4.3. The dependent variable is now a measure of patents instead of SBIR Phase II awards.[41] Table 21 defines the variables used in the patent estimation. As with Phase II awards, two variables are constructed to indicate innovative activity: PATDUM and PATENT. The variable, PATDUM, is a zero-one dummy variable that indicates whether or not a metropolitan area experienced some level of patent activity related to a given industry. The variable, PATENT, indicates the number of patents associated with a given industry in metropolitan areas that have had patent activity.

The patent data were compiled at the metropolitan area and industry levels using the PATSIC and MSA_ORI files available from the U.S. Patent and Trademark Office (USPTO). The PATSIC file includes SIC information for utility patents granted between 1963 and 1999. The MSA_ORI file associates a metropolitan area with every utility patent granted between 1990 and 1999 that has a first-named inventor residing in the U.S. The Technology Assessment and Forecasting Branch of the USPTO created a concordance linking USPTO patent classes and SIC codes in the mid-1970s. The concordance links patent classes to 41 specific SIC industries, which are listed in Appendix E. Patents were linked to the industries expected to produce the patented invention or use the invention in production (Griliches 1990). Patents, therefore, could be assigned to multiple industries. The methodology used for the concordance has been criticized because of double counting due to multiple SIC links and arbitrary links between patent subclasses and SIC categories (Scherer 1982a; Soete 1983).[42] However, few alternative methods are readily available in the scale of the PATSIC file.

Because of the nature of the USPTO data, this analysis is limited to utility patents, which are awarded for inventions. Types of patents excluded are plant patents, design patents, statutory invention registration documents, and defensive publications. These exclusions are not especially troublesome to this analysis given that most commercializable innovative activity is classified as an invention. The geographic location of patents is based on location of the first-named inventor and not the assignee that holds the rights to the patent. This method for determining location is expected to make little difference at the metropolitan area level for corporate patents, given that the inventor is usually an employee of the firm that is assignee and both are likely located in the same metropolitan area. There should also be no effect in this regard for individual patents where the inventor and assignee are the same.

Table 21. Definition of Variables for Patent Estimation

$POPDEN_s$	Average number of persons per square kilometer in a metro area, 1990-95
$R\&DLABS_s$	Average number of R&D labs located within a metro area, 1990-95
$EMPSIC73_s$	Average employment level in business services (SIC 73) within a metro area, 1990-95
$EMPCON_{is}$	Average location quotient for employment in industry i within a metro area, 1990-95
$UNIVDUM_s$	Dummy variable indicating whether (=1) or not (=0) any Research I/II or Doctorate I/II universities were located in a metro area, 1990-95
$UNIVR\&D_{is}$	Total level of academic R&D expenditures by Research I/II or Doctorate I/II institutions in fields corresponding to industry i within a metro area, 1990-95 (in thousands of 1992 Dollars)
$PATDUM_{is}$	Dummy variable indicating whether (=1) or not (=0) a metro area had any firms or individuals receive any utility patents in industry i, 1990-95
$PATENT_{is}$	Total number of utility patents received by firms and individuals in industry i within a metro area, 1990-95

The patent data are restricted to create a sample comparable to the SBIR Phase II data. Patents included in the analysis are those issued between 1990 and 1995 and those that have a unique two-digit industrial classification (to avoid double counting). The sample is also restricted to patents assigned to only one metropolitan area, eliminating patents with ambiguous locations. In addition, the patent sample includes only patents assigned to U.S. nongovernmental organizations and individuals to isolate patent activity in the private sector. Therefore, unlike the SBIR sample, the patent sample includes firms of any size as well as individual inventors.

The patent analysis controls for differential innovative activity across industries in the same manner as the SBIR analysis. However, the patent sample covers four of the five high-tech industries examined in Chapter 5: chemicals and allied products (SIC 28), industrial machinery (SIC 35), electronics and electrical equipment (SIC 36), and instruments (SIC 38). The PATSIC file does not classify SIC 87 due to the limited scope of the USPTO concordance between patent classifications and the SIC (see Appendix E). Therefore, analysis of this industry cannot take place using the patent sample.

Unlike the SBIR Phase II analysis in Chapter 5, the patent analysis employs a hurdle model without correction for potential sample selection. The patent model excludes correction for potential sample selection given the nature of patenting. The logic for this exclusion is that patenting requires an established research infrastructure so that the estimation of the unconditional model for patent counts is of little interest, in contrast to Phase II counts where SBIR activity can likely occur in a much less structured environment that is also influenced by other factors related to the metropolitan area and the SBIR Program.[43] Moreover, empirical estimation of the negative binomial patent equation controlling for selectivity indicates that selection is statistically insignificant and that the coefficients are similar to those from the estimation without selection correction.[44]

Table 22. Descriptive Statistics for Patents by Metropolitan Area, 1990-95 (N=273)

Variable	Mean	Std. Dev.	Minimum	Maximum
$PATENT_{28}$	32.8	1405	0	1,866
$PATENT_{35}$	39.0	107.0	0	870
$PATENT_{36}$	41.8	144.4	0	1,402
$PATENT_{38}$	24.5	86.6	0	736

Table 22 shows the means for the number of patents associated with a metropolitan area across the four industries in 1990-95. Instruments has the lowest number of patents, with an average of less than 25 patents in a metropolitan area. The maximum number of instruments patents is 736, close to half as much as the highest number in chemicals and electronics. Chemicals has a mean of approximately 33 patents but has the largest number of patents in a metropolitan area (1,866) across all four industries. The electronics industry has on average the highest number of patents (41) in a metropolitan area, suggesting electronics patent activity is more prevalent across metro areas than patenting in other industries or that several metropolitan areas have a relatively high rate of patent activity in electronics compared to other industries, which pulls up the industry average.

Tables 23 and 24 report descriptive statistics for the variables used in the patent model. Table 23 focuses on the small sample of metropolitan areas with no patent activity in the four industries during 1990-95, while Table 24 focuses on the metropolitan areas having some level of patent activity. The sixteen metropolitan areas with no patent activity look distinctly different than those with patents. These areas are characterized by weak technological infrastructures. They are not densely populated, have virtually no R&D labs, and have low levels of business services.

Table 23. Descriptive Statistics for Metropolitan Areas
Receiving No Utility Patents, 1990-95 (N=16)

Variable	Mean	Std. Dev.	Minimum	Maximum
POPDEN	42.2	27.4	4.5	121.1
R&DLABS	1.2	1.4	0	4.8
EMPSIC73	1,412.9	563.9	418.8	2,784.3
$EMPCON_{28}$	96.5	27.8	0	941.4
$EMPCON_{35}$	49.3	58.7	7.4	244.0
$EMPCON_{36}$	82.2	132.9	0.4	512.0
$EMPCON_{38}$	34.0	64.6	0	262.9
$EMPCON_{87}$	54.1	25.0	0	99.0
UNIVDUM	0.06	0.3	0	1.0
$UNIVR\&D_{28}$	832.8	3,331.4	0	13,325.5
$UNIVR\&D_{35}$	242.7	970.9	0	3,883.5
$UNIVR\&D_{36}$	96.5	386.1	0	1,544.4
$UNIVR\&D_{38}$	973.7	3,894.8	0	15,579.0
$UNIVR\&D_{87}$	1,091.7	4,366.7	0	17,466.6
$PH2_{28}$	0	0	0	0
$PH2_{35}$	0.1	0.3	0	1.0
$PH2_{36}$	0	0	0	0
$Ph2_{38}$	0	0	0	0
$Ph2_{87}$	0.1	0.5	0	2.0
$PATENT_{28}$	0	0	0	0
$PATENT_{35}$	0	0	0	0
$PATENT_{36}$	0	0	0	0
$PATENT_{87}$	0	0	0	0

Table 24. Descriptive Statistics for Metropolitan Areas
Receiving Utility Patents, 1990-95 (N=257)

Variable	Mean	Std. Dev.	Minimum	Maximum
POPDEN	104.8	105.8	1.9	954.0
R&DLABS	39.0	122.0	0	1,322.8
EMPSIC73	20,222.8	49,875.9	211.1	402,672.7
$EMPCON_{28}$	122.6	218.7	0.1	1,555.8
$EMPCON_{35}$	107.4	104.1	1.3	694.3
$EMPCON_{36}$	105.8	150.2	0	1,573.0
$EMPCON_{38}$	85.0	137.1	0	1,049.9
$EMPCON_{87}$	81.0	55.9	0	508.9
UNIVDUM	0.32	0.5	0	1.0
$UNIVR\&D_{28}$	140,612.3	475,280.9	0	4,062,545.1
$UNIVR\&D_{35}$	65,214.4	237,401.8	0	2,347,983.4
$UNIVR\&D_{36}$	40,236.1	166,005.2	0	1,799,800.1
$UNIVR\&D_{38}$	187,105.4	642,166.6	0	5,025,893.3
$UNIVR\&D_{87}$	249,192.3	800,420.6	0	6,163,377.4
$PH2_{28}$	1.5	6.0	0	51.0
$PH2_{35}$	1.1	4.5	0	46.0
$PH2_{36}$	2.8	12.8	0	132.0
$Ph2_{38}$	3.7	14.9	0	138.0
$Ph2_{87}$	6.9	31.2	0	371.0
$PATENT_{28}$	34.8	144.6	0	1,866.0
$PATENT_{35}$	41.4	109.8	0	870.0
$PATENT_{36}$	44.4	148.5	0	1,402.0
$PATENT_{87}$	26.1	89.0	0	736.0

Table 25. Difference in the Means for
Metropolitan Areas by Utility Patent Activity, 1990-95

Variable	Metro Areas With No Utility Patents (N=16) Mean	Metro Areas With Utility Patents (N=257) Mean
POP**	137,482.4	787,289.5
PDEN**	4.2	104.8
R&DLABS**	1.2	39.0
EMPSIC73**	1,412.9	20,222.8
$EMPCON_{28}$	96.5	122.6
$EMPCON_{35}$**	49.3	107.4
$EMPCON_{36}$	82.2	105.8
$EMPCON_{38}$**	34.0	85.0
$EMPCON_{87}$**	54.1	81.0
UNIVDUM**	0.06	0.32
$UNIVR\&D_{28}$**	832.8	140,612.3
$UNIVR\&D_{35}$**	242.7	65,214.4
$UNIVR\&D_{36}$**	96.5	40,236.1
$UNIVR\&D_{38}$**	973.7	187,105.4
$UNIVR\&D_{87}$**	1,091.7	249,192.3
$PH2_{28}$**	0	1.5
$PH2_{35}$**	0.1	1.1
$PH2_{36}$**	0	2.8
$PH2_{38}$**	0	3.7
$PH2_{87}$**	0.1	6.9
$PATENT_{28}$**	3.8	61.6
$PATENT_{35}$**	7.8	70.0
$PATENT_{36}$**	5.1	78.2
$PATENT_{38}$**	2.4	46.5

*Difference in the means is significant at the five percent level.
**Difference in the means is significant at the one percent level.

The concentration of industrial employment in these areas, on average, is considerably less than in the United States as a whole, particularly in instruments, machinery, and research services. Moreover, only six percent of these metropolitan areas have a research university located in the area, and in the areas with research universities, academic R&D activity is low. These same metropolitan areas are also absent of SBIR Phase II activity in the chemicals, electronics, and instruments industries and have almost no SBIR activity in machinery and research services.

The technological infrastructure in metropolitan areas with patent activity is strikingly different than that for the few metropolitan areas without patents, as evidenced by Table 25, which shows the difference in the means between these two groups of areas. The 257 metropolitan areas with patent activity have stronger technological infrastructures than those areas without patents. These areas on average are considerably more densely populated, have a much larger pool of business services, and have more concentrated employment across all four industries. Particularly noticeable is the stronger prevalence of research universities in these areas, with approximately 32 percent of these areas having one or more research universities located in the area. The level of R&D activity at universities is also considerably stronger. The metropolitan areas with patent activity also exhibit greater SBIR Phase II activity, suggesting that these areas have higher levels of different types of innovative activity than areas with no patents.

GEOGRAPHIC CONCENTRATION OF HIGH-TECH PATENT ACTIVITY

Table 26 lists the top five metropolitan areas that received patents in 1990-95 across four major industries. Looking at the four industries combined, New York received the most patents (7,674), followed by San Francisco, Boston, Los Angeles, and Chicago. Four of the top five metropolitan areas are found in coastal regions— two in California and two in the Northeast. Across the four industries, New York dominates in terms of patents received, being ranked first in chemicals and machinery and second in electronics and instruments. San Francisco and Boston are also ranked among the top metropolitan areas in all four industries. Chicago is the only non-coastal urban area with large numbers of patents across industries, evidenced by its being ranked among the top five metro areas in all but instruments.

The top metropolitan areas for patents closely resemble the top areas receiving Phase II awards (as seen in Table 5), though the rank order varies across the four industries. The major difference between patents and Phase II awards in terms of rankings is the presence of Chicago in the top list for patents and of Washington, DC, for Phase II awards. What is also noticeable between the highest ranked areas for patents and Phase II awards is the much lower concentration of patents in the top five metropolitan areas. Only 33 to 47 percent of all patents in 1990-95 were issued within the top five metropolitan areas compared to 50 to 58 percent of Phase II awards. This suggests that proximity at the metropolitan area level may not play as great a role in patent activity as in SBIR activity since geographic clustering within these industries is not as extreme for patents as for Phase II awards. However, the fact that at least a third of all patents in 1990-95 occurred within five metro areas is not insignificant. Clusters exist; they are just less concentrated than in the case of SBIR Phase II activity.

Table 26. Top Five Metropolitan Areas Receiving Utility Patents by Industry, 1990-95
(Number of Patents)

Four Industries Combined	Chemicals & Allied Products (SIC 28)	Industrial Machinery (SIC 35)	Electronics (SIC 36)	Instruments (SIC 38)
New York (7,674)	New York (1,886)	New York (870)	San Francisco (1,402)	Rochester (736)
San Francisco (3,270)	Philadelphia (992)	San Francisco (848)	New York (1,254)	New York (684)
Boston (2,160)	San Francisco (520)	Detroit (640)	Boston (729)	Los Angeles (545)
Los Angeles (1,992)	Chicago (471)	Boston (590)	Los Angeles (725)	San Francisco (500)
Chicago (1,929)	Boston (357)	Chicago (566)	Chicago (686)	Boston (486)
Percent of all Utility Patents Received by the Top Five Metro Areas				
37.2%	47.2%	33.0%	42.1%	44.0%

THE LIKELIHOOD OF PATENTING

Table 27 shows the empirical results for the probit step of the hurdle model described in Chapter 4 for four industries where the dependent variable is whether or not one or more utility patents have been issued to the private sector in a metropolitan area. The results indicate that knowledge spillovers as well as agglomeration effects influence patent activity. The number of R&D labs (R&DLABS) is positively and significantly related to the likelihood of a metropolitan area receiving one or more patents across all four industries. The concentration of industrial employment (EMPCON) is highly significant across each industry, suggesting that metropolitan areas with relatively higher employment in these industries are more likely to receive patents related to these same industries. Whether or not a metropolitan area receives any patents related to machinery, electronics, or instruments also depends on the availability of business services within that area, as measured by employment in business services (EMPSIC73). This evidence for significant agglomerative economies supports similar findings by Anselin et al. (1998) and Feldman (1994b) that demonstrate the positive effect of clusters of business services and industry employment on innovation counts. In contrast to these results indicating significant agglomerative economies, population density (POPDEN) has no effect on the likelihood of patenting in three of the four industries.   Only in the case of instruments does population density have a significant effect at the metropolitan area level.

Table 27. Probit Equation for the Likelihood of Patenting

Variable	Chemicals & Allied Products	Estimated Coefficients (Standard Errors) Industrial Machinery	Electronics	Instruments
Constant	-0.6606 (0.1762)	-0.6747 (0.3133)	-0.9624 (0.2364)	-0.9382 (0.2570)
POPDEN	-0.05333 (0.01441)	0.04628 (0.04329)	0.05154 (0.03271)	0.06538* (0.03729)
R&DLABS	0.07719*** (0.2186)	0.09011** (0.04322)	0.08735*** (0.02907)	0.1174*** (0.03812)
EMPSIC73	0.4905 (0.03107)	0.2374** (0.09307)	0.08293* (0.04431)	0.09567* (0.05137)
UNIVDUM	0.7284** (0.2946)	-0.2713 (0.3577)	0.1544 (0.3217)	0.7100* (0.3742)
EMPCON$_i$	0.01943*** (0.005642)	0.05041*** (0.01939)	0.02841*** (0.009977)	0.02714** (0.01161)
Ln L	-113.4	-70.7	-106.3	-94.6
χ^2	115.0	75.2	112.1	110.5

*Significant at the 10 percent level
**Significant at the 5 percent level
***Significant at the 1 percent level

What is strikingly absent from the likelihood of receiving patents is a strong effect of the presence of local research universities (UNIVDUM) across all industries.   While the presence of research universities increases the likelihood of receiving patents in chemicals and instruments, it has no significant effect whatsoever on the receipt of patents in the machinery and electronics industries.

This suggests that in the innovation process the benefit from proximity to research universities depends on the nature of the industry.

THE RATE OF PATENTING

The previous section provides evidence that agglomerative economies and local knowledge spillovers play a role in whether or not at least one industry-specific patent is issued within a metropolitan area. It remains to be seen if these agglomeration and spillover effects also influence the frequency of patent activity. To address this question, Table 28 shows the estimated effects of the local technological infrastructure on the number of patents received within a metropolitan area having patent activity by industry.[45] It is clear from the results that the positive impact of local knowledge spillovers and agglomeration on patent activity is strong.

Noticeable differences exist between the impact of the local technological infrastructure on the number of patents in areas with patent activity compared to the likelihood of patent activity across all metropolitan areas. The positive and significant role that R&D labs play in the first-stage probit estimations virtually disappears in the second stage. An increased presence of overall industrial R&D activity, as measured by the number of local R&D labs, has no significant effect on the number of patents in three of the four industries—chemicals, electronics, and instruments. Surprisingly, there is a significant, albeit weak, negative effect in machinery, which suggests that an increase in local R&D labs is associated with a small reduction in the number of patents related to the machinery industry.

University spillovers are more important in determining the number of patents than the likelihood of one or more patents being issued in a metropolitan area. In contrast to the probit results where the presence of research universities has a significant effect in only chemicals and instruments, an increase in university R&D expenditures leads to a significant increase in the number of patents issued within a metropolitan area in three of the four industries. The significant effect of university spillovers remains in chemicals. There is also a strong effect of university R&D activity on the frequency of patent activity in machinery, as well as a weaker—though still significant—effect in electronics. This suggests that the rate of patent activity benefits from nearby research universities through industry-related R&D activity in these two industries, though the likelihood of patenting in these industries is unaffected by merely the presence of these universities. The opposite is true in the instruments industry. The significant effect of the presence of research universities seen in the probit equation disappears when looking at the number of patents. Academic R&D expenditures do not play a significant role in the number of instruments-related patents in areas that experience some level of patent activity.

The size of the metropolitan area, as measured by population density (POPDEN), plays a drastically different role in whether or not patent activity occurs versus the number of patents if activity takes place. From the probit equations it is seen that population density has an insignificant effect on the likelihood of patent activity in each of the industries except instruments. For instruments, population density has a positive and modestly significant effect. In marked contrast, for all four industries, size is a strong determinant of the number of patents in areas with some level of patent activity. As the population grows denser, the number of patents significantly increases. This suggests that agglomerative economies indicated by the size of a metropolitan area play a positive and highly significant role in the intensity of patent activity at the metropolitan area level across all four industries that comprise most of

Table 28. Negative Binomial Equation for the Number of Patents

	Estimated Coefficients (Standard Errors)			
Variable	Chemicals & Allied Products	Industrial Machinery	Electronics	Instruments
Constant	1.2330	1.6180	1.8808	1.1331
	(0.2018)	(0.1721)	(0.1767)	(0.1319)
POPDEN	0.07945***	0.05185***	0.03495***	0.04284***
	(0.01360)	(0.008408)	(0.01171)	(0.007906)
R&DLABS	0.001892	-0.002571*	-0.002271	-0.0006587
	(0.001985)	(0.001445)	(0.001986)	(0.001408)
EMPSIC73	0.004672	0.02025***	0.02118***	0.01868***
	(0.004256)	(0.003417)	(0.004437)	(0.003019)
$UNIVR\&D_i$	0.005673**	0.01304***	0.01235*	0.001261
	(0.002350)	(0.003033)	(0.007211)	(0.001247)
$EMPCON_i$	0.02317***	0.03598***	0.02989***	0.03223***
	(0.003246)	(0.009126)	(0.005632)	(0.003776)
α	1.1563***	0.9704***	1.3663***	0.8436***
	(0.1502)	(0.09818)	(0.1775)	(0.1018)
Ln L	-719.6	-954.3	-814.4	-717.0

*Significant at the 10 percent level
**Significant at the 5 percent level
***Significant at the 1 percent level

the high-tech sector. The strong positive effect of the size of a metropolitan area seen here is consistent with a similar effect found by Jaffe (1989) and Feldman (1994b) at the state level.

In a similar vein, agglomerative economies resulting from the availability of business services also have a highly significant and positive impact on the intensity of patent activity. Growth in business services employment (EMPSIC73) contributes to an increase in the number of patents issued within a metropolitan area. The level of employment in business services has a similar effect on both the likelihood and intensity of patent activity, though the effect is more significant in relation to the number of patents. In both cases, the positive and significant effect of business services occurs in the same three industries—machinery, electronics, and instruments. This suggests these industries rely more heavily on business services in the innovation process than does the chemicals industry.

Along with population density as a measure of the size of a metropolitan area, the employment concentration in a given industry (EMPCON) is the only component of the technological infrastructure that significantly affects the number of patents across all four industries. Industrial employment concentration exhibits a highly significant and positive effect on the intensity of patenting in areas with patent activity, similar to its effect on whether or not one or more patents are issued within a metropolitan area. Therefore, metropolitan areas with a greater employment concentration in a given industry than the United States as a whole display significantly greater numbers of patents compared to areas with relatively low or equivalent employment concentrations.

COMPARISON TO SBIR PHASE II ACTIVITY

Source of Agglomeration and Spillover Effects

Tables 29 and 30 compare the probit and negative binomial equations for Phase II awards and patents. Focusing on levels of significance, these results suggest that geography affects patent activity through both agglomerative economies and knowledge spillovers, whereas in the case of Phase II activity the effects of geography are more restricted to knowledge sources than agglomeration sources. Taken together, the results for Phase II awards and patents suggest that geographic proximity matters in the innovative process but that the effect of knowledge spillovers and agglomeration play a different role based on the type of innovative activity that is pursued. Of course, this could be due (at least in part) to the inclusion of large firms in the patent data, which cannot be directly examined in this analysis. It should also be recalled that the negative binomial results for patents and Phase II awards were estimated differently, with the patent equation conditional on areas with patent activity and the Phase II award equation being unconditional for all metropolitan areas. This limits direct comparison of the negative binomial results for patents and Phase II awards.

With regard to specific variables, the results are similar in some ways and strikingly different in others between patents and Phase II awards. The importance of proximity to R&D labs, for example, follows a similar pattern in its significant effect on the likelihood of Phase II awards and patents, as well as its lack of an effect on the number of awards and patents. The importance of industrial R&D to the likelihood of innovative activity is in line with previous findings that indicate industrial R&D is significantly related to innovative activity (Anselin et al. 1997, 2000; Acs et al. 1992, 1994; Feldman 1994b; Jaffe 1989). The finding that the rate of innovation is not related to industrial R&D activity, however, is not in line with these findings. This differential may be related to the failure of previous researchers to use a two-stage estimation strategy to investigate the importance of geography. It may also be due to using a smaller unit of observation—the city—as opposed to the state or to the aggregation of the R&D labs variable across industries in this analysis.

The almost nonexistent effect of an area's size (POPDEN) seen in the likelihood of Phase II awards is carried over to the likelihood of patents, suggesting that agglomerative economies from the size of a metropolitan area have little effect on the likelihood of innovative activity across most high-tech industries.

Compared to the Phase II award results, however, the patent findings suggest that agglomerative economies play a larger role in patenting than in SBIR activity. The agglomeration effects are more noticeable in relation to the rate of innovation than the likelihood of innovation. Population density is significant and positive for patents in all four industries, while it is only positive in machinery for Phase II awards. Business services have a significant effect on the number of patents and no effect on Phase II awards across all four industries. Industrial employment has a positive and significant effect on the rate of patenting across all four industries but no impact whatsoever on the number of Phase II awards in three of the four industries. In the remaining industry (chemicals), the effect is negative for Phase II awards. These results suggest that geographic proximity plays a more significant role on the frequency of patent activity than on SBIR Phase II activity, despite the lower geographic concentration of patents compared to Phase II awards. However, the differences in these results may be partly driven by the estimation of the negative binomial patent equation only for areas with patent activity, while the Phase II negative binomial equation was based on all metropolitan areas. It is plausible that agglomeration effects are more evident in areas with innovative activity than in all metropolitan areas regardless of innovative activity.

What is strikingly different is the weaker link between research universities and innovative activity as measured by patents instead of Phase II awards. Proximity to research universities has a significant effect on the likelihood of Phase II activity in all four industries, as well as on the number of awards in every industry but machinery. In contrast, university spillovers have a significant effect on the

Table 29. Comparison of SBIR Phase II Award and Patent Estimates for Probit Equation

	Chemicals & Allied Products		Industrial Machinery		Electronics		Instruments	
				Estimated Coefficients (Standard Errors)				
Variable	Phase II Awards	Patents	Phase II Awards	Patents	Phase II Awards	Patents	Phase II Awards	Patents
Constant	-1.8174 (0.2017)	-0.6606 (0.1762)	-1.6282 (0.2227)	-0.6747 (0.3133)	-1.7269 (0.1763)	-0.9624 (0.2364)	-1.7559 (0.1915)	-0.9382 (0.2570)
POPDEN	0.2229** (0.1031)	-0.005333 (0.01441)	-0.1985 (0.1823)	-0.04628 (0.04329)	-0.02726 (0.1236)	-0.05154 (0.03271)	-0.008945 (0.1482)	0.06538* (0.03729)
R&DLABS	2.1743*** (0.5718)	0.07719*** (0.2186)	0.9722** (0.5118)	0.09011** (0.04322)	1.9766*** (0.5531)	0.08735*** (0.02907)	1.9755*** (0.8226)	0.1174*** (0.03812)
EMPSIC73	-0.1566** (0.07819)	0.4905 (0.03107)	0.1441* (0.09459)	0.2374** (0.09307)	0.05241 (0.09401)	0.08293* (0.04431)	0.2360** (0.1323)	0.09567* (0.05137)
UNIVDUM	1.02763*** (0.2325)	0.7284** (0.2946)	0.5787** (0.2477)	-0.2713 (0.3577)	0.5994*** (0.2402)	0.1544 (0.3217)	0.7900*** (0.2350)	0.7100* (0.3742)
EMPCON$_i$	-0.03651 (0.07231)	0.01943*** (0.005642)	0.05087 (0.1139)	0.05041*** (0.01939)	0.08114 (0.06648)	0.02841*** (0.009977)	0.1831*** (0.07460)	0.02714** (0.01161)

*Significant at the 10 percent level
**Significant at the 5 percent level
***Significant at the 1 percent level

Table 30. Comparison of SBIR Phase II Award and Patent Estimates for Negative Binomial Equation[1]

Estimated Coefficients
(Standard Errors)

Variable	Chemicals & Allied Products		Industrial Machinery		Electronics		Instruments	
	Phase II Awards	Patents	Phase II Awards	Patents	Phase II Awards	Patents	Phase II Awards	Patents
Constant	1.9966 (0.3028)	1.2330 (0.2018)	1.1123 (0.5222)	1.6180 (0.1721)	2.3700 (0.4933)	1.8808 (0.1767)	2.08541 (0.3343)	1.1331 (0.1319)
POPDEN	-0.2816* (0.1550)	0.07945*** (0.01360)	0.3716* (0.2415)	0.05185*** (0.008408)	-0.1068 (0.1436)	0.03495*** (0.01171)	-0.1446* (0.08844)	0.04284*** (0.007906)
R&DLABS	0.2111 (0.1654)	0.001892 (0.001985)	0.01118 (0.2811)	-0.002571* (0.001445)	0.2908 (0.2391)	-0.002271 (0.001986)	0.1965 (0.1684)	-0.0006587 (0.001408)
EMPSIC73	0.01066 (0.03150)	0.004672 (0.004256)	-0.01349 (0.05721)	0.02025*** (0.003417)	-0.03806 (0.05195)	0.02118*** (0.004437)	0.01147 (0.04832)	0.01868*** (0.003019)
UNIVR&D$_i$	0.04108* (0.02498)	0.005673** (0.002350)	0.06474 (0.05228)	0.01304*** (0.003033)	0.1369** (0.06549)	0.01235* (0.007211)	0.03013* (0.02068)	0.001261 (0.001247)
EMPCON$_i$	-0.4179*** (0.1628)	0.02317*** (0.003246)	-0.2322 (0.2501)	0.03598*** (0.009126)	-0.05283 (0.1386)	0.02989*** (0.005632)	0.02696 (0.06346)	0.03223*** (0.003776)

[1]Direct comparison between the results for Phase II awards and patents is limited due to sample-selection in the Phase II estimations.
*Significant at the 10 percent level
**Significant at the 5 percent level
***Significant at the 1 percent level

likelihood of patenting only in chemicals and instruments. Academic R&D has a positive and significant effect on both the number of patents and Phase II awards in chemicals and electronics. A significant effect is only present for Phase II awards in instruments; the opposite is true for machinery. This differential effect of university spillovers may be due to the size of the firms involved in SBIR and patent activity. As previous evidence has indicated, small firms may benefit more than large firms from university research (Acs, Audretsch and Feldman 1992; Feldman 1994b). Because Phase II activity is restricted to small firms only and the measure of patent activity includes firms of any size, the weaker effect of universities on patent activity compared to Phase II activity is plausible. The difference may also be caused by differences in the type of innovative activity pursued and in the sample sizes for the second-stage negative binomial estimations. A patent can be considered a by-product of advanced research, while a Phase II award represents funding for in-progress research. A patent may be tied to research that benefited from proximity to universities earlier in the innovation process; SBIR research, on the other hand, may have more recent ties to local universities, suggesting that the relationship between university spillovers and Phase II awards would be stronger than for patents. In addition, the effect of university spillovers may appear weaker for patents because the patent count sample for the negative binomial estimation only includes metropolitan areas with some level of patent activity, whereas the Phase II award count sample includes all metropolitan areas. Regardless of these caveats, the evidence suggests that university spillovers are a necessary condition for most innovative activity.

Magnitude of Agglomeration and Spillover Effects

The previous results in this analysis have identified the sources of significant agglomerative economies and knowledge spillovers for Phase II awards and patents. They have not provided a direct measure of the magnitude of these effects on innovative activity. While it is important to understand how the local technological infrastructure influences innovative activity, it is also important to know the size of this influence. Knowing the magnitude of these effects offers policymakers useful information on determining specific targets for innovation policies. For instance, in evaluating funding alternatives, it would be informative to know the magnitude of the effect rather than simply the presence of an effect.

This analysis examines the estimated probability that innovative activity will occur in a metropolitan area with a given level of technological infrastructure as an indicator of the magnitude of agglomeration and spillovers effects. This measure is evaluated for two representative states of the local technological infrastructure. One state represents a 'weak' infrastructure and one a 'strong' infrastructure. The weak

Table 31. Probability of Innovative Activity in a Metropolitan Area

| | Chemicals & Allied Products | | Industrial Machinery | | Electronics | | Instruments | | Research Services | |
| | Infrastructure | | Infrastructure | | Infrastructure | | Infrastructure | | Infrastructure | |
	Weak	Strong	Weak	Strong	Weak	Strong	Weak	Strong	Weak	Strong
Phase II Awards	0.000	0.000	0.001	1.000	0.000	0.999	0.030	1.000	0.001	0.994
Patents	0.918	1.000	0.036	1.000	0.001	0.997	0.003	0.999	NA	NA

NAProbabilities were not calculated for SIC 87 because this industry cannot be identified in the patent data.

infrastructure is based on the average characteristics of the technological infrastructures for the metropolitan areas having no SBIR Phase II activity and represents a typical metropolitan area that has a relatively small population, little R&D activity, few if any research universities, and low concentrations of business services and industry-level employment (see Table 12). The strong infrastructure is based on the average characteristics for the metropolitan areas with some level of Phase II activity. In this case, the infrastructure is large and densely populated and has substantial levels of R&D activity and business services.

Table 31 presents the probabilities that Phase II activity and patenting by industry will take place in a metropolitan area based on the representative weak and strong infrastructures. It is strikingly clear that the likelihood of innovative activity depends on the strength of the local technological infrastructure. For machinery, electronics, instruments, and research services, Phase II activity is virtually guaranteed to occur in metropolitan areas having a strong technological infrastructure related to these industries. The opposite is true for metropolitan areas with weak infrastructures where the probability of Phase II activity is virtually zero for these industries. The probability of patent activity in machinery, electronics, and instruments mimics that of Phase II activity.

The probabilities for both Phase II and patent activity in the chemical industry are drastically different than those in the other industries. There is no difference between the weak and strong infrastructures for Phase II awards. In both cases, the predicted probability of Phase II activity is zero. At the other end of the spectrum, the probability is high that patent activity in the chemicals industry will occur in metropolitan areas with either a weak or strong infrastructure.

SUMMARY

The empirical findings presented in this chapter indicate that knowledge spillovers and agglomerative economies lead to greater patent activity. The presence of R&D labs, business services, and concentrated industry employment have a positive effect on the likelihood of patenting across all four industries, while the presence of research universities has a similar effect only in chemicals and instruments.

This chapter also provides evidence that the local technological infrastructure in many ways has a similar effect on innovative activity, measured either by patents or Phase II awards. However, its impact on these different measures of innovation varies largely in the effect of agglomerative economies and spillovers from business services and industrial employment. The existence and magnitude of spillovers from research universities also varies across industries between the two innovation measures.

The difference in the impact of the local technological infrastructure on Phase II and patent activity may be driven by three factors. First, small firms solely perform

SBIR activity, while firms of any size carry out patenting. Therefore, differential impacts may indicate differences in the innovation mechanisms between large and small firms. Second, the differences may be related to the type of innovative activity that the two measures capture. SBIR activities are often quite different from patent activity, and this difference may influence how firms appropriate available knowledge in these different types of innovative processes. Unfortunately, we cannot distinguish between these two reasons, given that the data do not permit disaggregation of patent counts by firm size. Third, direct comparison of the results for Phase II awards and patents is limited due to the use of the unconditional model for Phase II counts and the conditional model for patent counts in the second-step, negative binomial estimations. The estimated results, therefore, may not fully reflect differences between Phase II and patent activity, but to some extent, variation due to differences in the estimated models.

This chapter further provides estimates of the magnitude of the impact of the local technological infrastructure on innovative activity. The strength of the infrastructure has a large impact on the probability of innovative activity occurring, regardless of the measure of innovation being Phase II awards or patents. In all but the chemicals industry, the probability of innovative activity is virtually zero in areas with weak infrastructures, while the probability is essentially one in areas with strong infrastructures. At the very least this suggests that a strong infrastructure is necessary for innovative activity to take place. Chapter 8 will examine the implications of this study's empirical findings and their application to economic and public policy.

8
CONCLUSION

The empirical analysis presented in this book has endeavored to expand the understanding of the relationship between geography and the innovative activity of small firms. It introduced a novel measure of small firm innovation, the Phase II award from the Small Business Innovation Research Program. It has identified the existence and importance of localized knowledge spillovers and agglomeration—as well as funding agency effects—in the innovation process of high-tech small firms at the metropolitan area level using this unique measure of innovation. It has also examined the effect of the local technological infrastructure on patents and explored whether the infrastructure has a differential effect when patent counts by industry are used as the measure of innovation instead of Phase II awards.

This research extends the body of work on the role that geographic proximity plays in the innovation process in four meaningful ways. First, it uses a novel measure of innovation, SBIR Phase II awards, to test the impact of the local technological infrastructure on innovative activity. This measure expands the means by which innovative activity in the United States has been, and can be, examined. Second, it targets the small business sector, a growing contributor to economic activity that has received less attention in the innovation literature. Third, it provides an analysis of agglomeration and knowledge spillover effects at the metropolitan level as opposed to broader geographic areas, such as the state. Finally, by examining innovation during the 1990s, this work provides insight into whether the effects of geographic spillovers on innovation have changed since the 1980s, the period for which most previous research provided evidence.

SUMMARY

    The local technological infrastructure captures the effects of agglomeration and knowledge spillovers on innovative activity in a geographic area. Agglomerative economies can enhance local innovation through the concentrations of economic activity, relevant labor pools, and available business services important to the innovation process. Innovation may also be stimulated by knowledge spillovers originating from agents and institutions in both the private and public sectors. The most common sources of spillovers emanate from industrial R&D activity and university research. A growing body of literature has provided empirical evidence that indicates a significant impact of the local technological infrastructure (or at least elements of the infrastructure) on innovative activity.

    Building upon this body of literature, this research examined the role of geographic proximity in the innovation process of high-tech small firms in the 1990s using the Phase II award as a novel measure of innovation. This research has shown that the SBIR Phase II award is a useful measure of small firm innovation in the United States given its clear connection to innovation, focus on high-tech small firms, and availability of data. The SBIR Program is the largest federal R&D initiative targeting the small business sector and is designed to stimulate commercialized innovation. Phase II of the SBIR Program provides federal funding for research during the early stages of the innovation process and targets research that shows promise of commercialization. The Phase II award resembles a patent in that it is an intermediate outcome in the innovation process and does not measure an actual innovation, though it can be argued that Phase II research is more closely related to commercialized innovation than patents. Evidence indicates that Phase II research is clearly connected to commercialized innovation, with a substantial segment of Phase II research achieving market sales. The geographic distribution of Phase II activity follows a similar pattern as that of other innovation measures, including R&D expenditures and patenting. Phase II activity is concentrated in California and the Northeast—particularly in Boston, San Francisco, and New York—as are R&D expenditures and patents. As a useful measure of innovation, the SBIR Program provides centralized data on award winning firms across time, including firm location.

    A knowledge production function where industrial R&D activity, research universities, the concentration of industrial employment, the availability of business services, and population density are determinants of innovative activity was estimated. The empirical model was based on a negative binomial hurdle model to capture the separate effects of the local technological infrastructure on the likelihood of innovative activity and on the rate of innovation at the metropolitan level. The empirical estimation relies on a unique data set of SBIR activity during 1990-95, which required extensive effort to compile at the metropolitan and industry levels. The model was estimated separately for five industries comprising the high-technology sector: chemicals and allied products, industrial machinery, electronics, instruments, and research services.

The empirical findings for Phase II activity support the view that geography matters in the innovation process. Geographically bounded spillovers from the technological infrastructure are particularly instrumental in whether or not firms receive Phase II awards. Private- and public-sector knowledge spillovers, indicated by the number of R&D labs and the presence of research universities, significantly affect the likelihood of Phase II activity across all five industries. The significant and positive effect of research universities persists when explaining the number of Phase II awards in four of the five industries. However, there is no evidence of an impact of R&D labs on the number of Phase II awards.

Cross-industry agglomeration effects from the prevalence of business services and population density more clearly determine the likelihood of receiving Phase II awards than the rate of awards within a metropolitan area. There is also no consistent effect of industrial employment concentration on either the likelihood or number of Phase II awards. Employment concentration is positive and significant only for instruments and research services in explaining the likelihood of Phase II activity and for research services in explaining the number of awards.

These findings support the hypothesis that small firms tend to appropriate external sources of knowledge given internal resource constraints. They also suggest that knowledge spillovers play a more consistent role than agglomeration in determining the likelihood of innovative activity within a metropolitan area. However, agglomeration plays an increasing role and knowledge spillovers a lesser role in determining the rate of innovative activity. These results also suggest that research universities—the predominant source of basic research—are the most consistent factor in both the likelihood and frequency of Phase II activity at the metropolitan area level, suggesting that institutions of higher education are an important source of knowledge for small firms.

In addition to variation across industries in the effect of the local technological infrastructure, this research also finds that its effect depends to some extent on the government agency funding SBIR research. A distinct difference exists in the impact of the technological infrastructure on the likelihood of Phase II activity among the top three funding agencies, particularly between HHS and DOD. The effects on DOD and NASA funded Phase II activity are similar due to the frequently common nature of their research. The clearest difference between agencies centers on the importance of different types of knowledge to Phase II activity. Private knowledge, as measured by R&D labs, generally plays a significant role in Phase II activity funded by DOD and NASA but plays little or no part in HHS funded activity. The opposite is true for spillovers from universities, which have a more significant impact on HHS activity than on DOD or NASA activity.

It should be no surprise that differences arise across agencies. These results support the common perception that agencies operate under different spheres of connectedness. Agencies face different types of research agendas and attract different types of researchers, creating distinct networks between agencies and the research community. DOD, for instance, has traditionally forged relationships with industry, while HHS has stronger ties with the academic sector. Another notable difference between DOD and the other agencies is the importance of proximity to military installations in industrial machinery, instruments, and research services. This result can be attributed to the nature of defense-oriented research, which typically requires increased security, face-to-face interaction with military personnel, and frequent use of government facilities.

The question arises from the empirical findings for Phase II activity of whether these results are indicative of innovative activity in general or are driven by the structure of the SBIR Program. The usefulness of the Phase II award as a measure of innovation depends on its ability to provide insight into innovative activity beyond the boundaries of the SBIR Program. As a means of comparison, a comparable analysis was performed using patents as the measure of innovation to examine the impact of the local technological infrastructure on innovative activity. Patents have been the most common measure of innovation applied in previous research, and patent activity has been shown to benefit from geographically bounded spillovers at the state level. This analysis estimated the effect of agglomeration and knowledge spillovers on patent activity across four industries at the metropolitan level.

The results from the patent analysis support previous evidence concerning the importance of geographic proximity to patenting. As with Phase II activity, local spillovers and agglomerative economies matter for patenting at the metropolitan level. R&D labs and the presence of research universities have a significant impact on the likelihood of patent activity. Similar to Phase II activity, R&D labs generally have no effect on the number of patents, while the impact of university R&D activity on the number of patents is significant. The findings for the positive effect of R&D labs on the likelihood of innovative activity are consistent with previous research indicating a significant relationship between industrial R&D activity and innovation. The lack of an effect of R&D labs on the rate of innovation, however, is not in line with previous research. This may be due to the two-stage estimation technique, the smaller unit of observation (metropolitan areas) used in this analysis as compared to previous research, and/or the way R&D labs are aggregated across industries.

The most noticeable difference between Phase II activity and patenting are the effects of agglomeration. The concentration of industrial employment, availability of business services, and population density have a significant impact on patent activity, particularly the number of patents, across almost all four industries; their effects on Phase II activity are in some cases weaker or unseen altogether. The increased importance of agglomeration for patents compared to Phase II awards may be driven by the nature of patenting which requires an established research infrastructure and likely draws more heavily from services outside the firm, such as legal counsel and patent consultants, than SBIR activity.

The combined results for Phase II activity and patenting suggest that these differences in the role of the technological infrastructure on innovative activity may not be trivial. One reason for this differential impact is that different types of innovative activity may utilize the technological infrastructure differently. The innovation processes leading to Phase II activity are likely not the same as those for patenting. Therefore, the benefits from knowledge spillovers and agglomeration may vary by type of innovative activity. The differences may also be caused by variation in firm size. Phase II activity is by definition restricted to small firms only. Patenting, as measured here, includes firms of all sizes. The impact of firm size, however, was not analyzed because patents could not be disaggregated by firm size.

POLICY IMPLICATIONS

Two distinct veins of policy implications emerge from this research. The first centers on the SBIR Program or any similar innovation policy, and the other around economic development. Concerns about the geographic distribution of SBIR funding have escalated among SBIR legislators and administrators. Some question whether the highly skewed distribution of SBIR awards lines up with the goals of a program designed to stimulate innovation among small firms disproportionately overlooked by other federal R&D activities. Recent recommendations urge a more equitable distribution of awards at the state level and increased involvement of state and local government.

The research presented in this book supports evidence that agglomeration and knowledge spillovers contribute to the clustering of innovative activity, which likely drives the skewed distribution of SBIR awards. Previous research also indicates innovation is clustered in certain geographic areas, such as California or the Northeast, and that these areas have greater levels of SBIR activity. Inefficient outcomes from the awards-selection process may arise if policies designed to more equitably distribute SBIR awards lead to unsuccessful firms being chosen over successful ones. Innovative activity would diminish if these unsuccessful firms contribute less to economic activity than the successful firms they replace. SBIR activity funded in areas that lack sources of spillovers may also require augmented levels of resources to be successful. Attempts to create a more equitable distribution, therefore, potentially may reduce the effectiveness of the SBIR Program to fund successful projects—the 'ultimate goal.' Such policies could also interfere with local and state economic development initiatives that increasingly include the SBIR Program or other small business initiatives.

Although this research does not provide direct evidence on the contributions of SBIR research to economic growth, it does suggest that Phase II activity is a useful measure of innovative activity. Phase II activity is closely linked to innovation generating commercial sales and resembles other measures of innovation. Given this evidence, policymakers may well consider the SBIR Program as a potential avenue for stimulating innovation among the small business sector in their areas. However, they should exercise caution in automatically assuming SBIR research results in significant economic gains for their local economies. Further research is needed to estimate the economic impact of the SBIR Program at the local level.

By reaffirming that the local technological infrastructure plays a role in stimulating innovation among small firms, this research provides useful information to policymakers interested in boosting economic activity within metropolitan areas. Concentrated efforts by state and local policymakers to create regional hotspots of innovation are not imaginary. For example, at BIO 2001, the largest convention in the biotechnology industry, representatives from almost every U.S. state and 40 countries were on hand to recruit biotech companies to locate in their regions (Keefe 2001). Similarly, in 2001, the State Science and Technology Institute hosted its fifth annual conference on creating high-tech local economies, including discussions on increasing the role of universities and utilizing federal science and technology programs (Southern Growth Policies Board 2001). Policymakers, therefore, must be well informed about what drives innovative activity at the state and local levels.

The empirical results of this research have several implications related to economic development policies. First, local economic development policies should not ignore the small business sector. Innovative activity by small firms, as measured by Phase II activity, is prevalent to some degree in approximately half of U.S. metropolitan areas. Second, the larger the technological infrastructure, the more likely a metropolitan area will experience innovative activity. Incentives to attract innovative firms may fall short if an infrastructure providing sufficient agglomerative economies and knowledge spillovers is not in place. Early development efforts would perhaps be better focused on building a suitable technological infrastructure, such as establishing private and public research facilities. Subsequent policies could then be directed to stimulate the flow of knowledge between agents and institutions in the local economy. Incentives used to stimulate innovation may include R&D tax credits, corporate tax reductions, targeted funding for education, and government programs to aid small and new firms in the innovation process, such as firm incubators.

Third, economic development policies must take into account the state of the current technological infrastructure and the goals of policymakers related to the type of development desired. It is not enough to only know that the technological infrastructure can stimulate innovative activity. For policy to be most effective, it must also draw on the strengths of the current infrastructure, address the infrastructure's weaknesses, and target specific types of innovative activity. Georgia, for instance, has been working to lure biotech companies for over three years and has increased its efforts by recently hiring a full-time biotech recruiter in the Department of Industry, Trade and Tourism and by commissioning a study of the state's strengths and weaknesses as a location for biotech firms (Keefe 2001). As a way to attract biotech firms, Georgia is considering creating a seed-capital fund for early-stage firms and tax incentives. Georgia's efforts are paying off as the number of biotech companies in Georgia has risen by approximately 20 new firms in one year alone, raising the total number of firms to around 90 and increasing the cluster of biotech firms around Atlanta (Keefe 2001; Walcott 1999).

The differential effects of the technological infrastructure across SBIR funding agencies means that policymakers should also consider the types of publicly funded research likely to be performed in their areas and target their policies accordingly. DOD and HHS Phase II research rely on components of the technological infrastructure in varying degrees. These differences likely arise from each agency having a different sphere of connectedness, leading agencies to interact with different sources of knowledge spillovers within the technological infrastructure, such as private firms or universities. Policymakers, therefore, should target policies to enhance interaction between local researchers and agencies that would likely fund research in their areas.

Development policies should strive to enhance elements of the technological infrastructure that most benefit publicly funded research in their areas. For example, areas that desire to concentrate on biotech-oriented research that would be supported by HHS should target the creation of a strong academic sector with expertise in the biosciences. Legislative focus on the role of universities as a tool for economic development is evident in a recent survey of selected state legislators from all 50 U.S. states (Ruppert 2001). Legislators believe higher education is vital to economic development and that economic development initiatives drive funding agendas for higher education. These legislators agree that universities stimulate local economic activity and attract new firms by providing knowledge and ideas and by generating a high-skilled workforce. As a result, states are targeting education funds increasingly towards programs that generate a higher-skilled workforce. State spending for student aid increased almost 30 percent between 1994-1995 and 1999-2000, and merit-based scholarship programs accounted for 22 percent of this funding.

FUTURE DIRECTIONS

The body of work in this book by no means fully answers questions about the relationship between geography and small-firm innovation but opens the door for further research in several directions. Additional research could overcome several limitations of the analyses presented here. One of the most significant barriers to the application of SBIR data to the study of innovation is the difficulty in linking SBIR firms to industrial classifications. Linking the SBIR data to additional firm-level databases, such as Compustat, could provide further insight on the firms performing SBIR research. Furthermore, the low sample sizes for Phase II data in certain industries, which limit empirical estimation, could be augmented by extending the period of study to include more recent years of Phase II activity, by combining Phase I and Phase II awards, or by broadening the geographic unit of observation.[46] Moreover, a similar analysis at the state level would also allow for the addition of industrial R&D expenditures to the model, which could arguably be a better indicator of the creation of industrial knowledge than R&D labs. While this analysis provides evidence of the "average" effect of the local technological infrastructure on innovative activity over a six-year period, it does not shed light on whether this effect has changed over time. An insightful extension of this research would explore this time dimension by exploiting the time series nature of the SBIR data.

Second, the empirical results prompt further examination of the variation in the effect of the local technological infrastructure across industries and funding agencies. Research focusing on a larger range of industries, including those not necessarily knowledge-oriented, would provide more conclusive evidence on the role of knowledge spillovers and agglomeration in the innovation process. Focusing on more narrowly defined industries (in contrast to the two-digit industries examined in this research) would allow researchers to isolate the effects of the technological infrastructure within small sectors of industrial activity. This would be particularly useful for policymakers since it is expected that proximity plays a more important role as technological space grows closer.

Evidence from this research also suggests that factors associated with the agency funding SBIR research influence the impact of the technological infrastructure on Phase II activity. It has been argued that this differential impact emerges from differences in the spheres of connectedness that shape agencies' research agendas and the types of researchers seeking funding. An important line of future research could more fully examine the role of agency characteristics in publicly funded innovation.

Third, much research remains to be done to examine the policy implications resulting from the relationship between geography and innovation. This work provides a detailed analysis of the geographic concentration of SBIR Phase II activity. Policymakers would benefit from future research that examines the economic impact of current and proposed policies on the geographic distribution of SBIR awards, which has gained increasing attention in recent years. Furthermore, the empirical evidence presented here suggests that public policy may have a role in strengthening the impact of the local technological infrastructure on innovation by increasing the flow of knowledge between agents and institutions and by stimulating the clustering of these agents and institutions in local areas. Further research is needed to examine the most efficient structure of innovation policies for specific locations. For instance, metropolitan areas with different states of their technological infrastructures and different goals of economic development should likely rely on different combinations of incentives to stimulate innovative activity.

CONCLUSIONS

Three major conclusions can be drawn from this research. First, the Phase II award from the Small Business Innovation Research Program is a meaningful measure of innovative activity for the high-tech small business sector. It is particularly useful to examine publicly funded innovation given that SBIR research is in part or wholly funded by federal government agencies and this funding is well documented. The Phase II award is appealing as a measure of innovation in that it is closely related to commercialized innovation, provides a measure of small-firm innovative activity, resembles other measures of innovation in its geographic distribution, and allows for analysis over time.

Second, the empirical evidence indicates that the impact of agglomeration and knowledge spillovers is not constant across geographic regions. Variation can exist across industries, type of innovative activity, firm size, and government agencies involved in funding research. Moreover, the technological infrastructure affects the likelihood of innovative activity and the rate of innovation differently. Knowledge spillovers play a greater role than agglomeration economies in Phase II activity at the metropolitan level, while the effects of agglomeration take on a more significant role in patenting. Of particular interest is the importance of local university spillovers to innovation, particularly the level of innovative activity. These findings reinforce previous research indicating that geographic proximity to spillovers and agglomeration matter to the innovation process and are consistent with other research that identifies the importance of knowledge spillovers to small-firm innovation.

Third, this research proves useful for innovation and economic development policies. It suggests that the skewed concentration of SBIR activity stems from the presence of knowledge spillovers and agglomerative economies. Rash implementation of policies based on geographic equity may thus have the unintended consequence of reducing the efficiency of the SBIR Program. Effective policies to stimulate local economic development must consider the composition and role of the local technological infrastructure in the innovation process. Policymakers, therefore, must understand the nature of research, the sources of agglomerative economies and knowledge spillovers, and the means of interaction between agents and institutions within their region. Good policy cannot ignore geography's impact on innovation.

APPENDIX A

INDUSTRIES COMPRISING THE HIGH TECHNOLOGY SECTOR

There is no standardized definition of what constitutes 'high technology.' For a review of the most common definitions, see DeVol (1999), Hadlock et al. (1991), Hecker et al. (1999), Luker et al. (1997), National Science Board (1998), and Walcott (2000). The high-technology sector is defined in this study as industries classified within five major industrial groups: chemicals and allied products, industrial machinery, electronics and electrical equipment, scientific instruments, and research-oriented services. The 1987 Standard Industrial Classification (SIC) system for the United States is used to define industries. Following this system, the high-technology sector includes industries classified under SIC Major Industry Group codes 28 (Chemicals and Allied Products), 35 (Industrial Machinery & Equipment), 36 (Electronic & Other Electrical Equipment), 38 (Instruments and Related Products) and 87 (Scientific and Management Services). The industries at the four-digit SIC level that comprise each of these Major Industry Groups are listed below.

SIC 28 - CHEMICALS AND ALLIED PRODUCTS

SIC
Industry
Number   *Industry*

2812	Alkalies and Chlorine
2813	Industrial Gases
2816	Inorganic Pigments
2819	Industrial Inorganic Chemicals, Not Elsewhere Classified
2821	Plastics Materials, Synthetic Resins, and Nonvulcanizable Elastomers
2822	Synthetic Rubber (Vulcanizable Elastomers)
2823	Cellulosic Manmade Fibers
2824	Manmade Organic Fibers, Except Cellulosic
2833	Medicinal Chemicals and Botanical Products
2834	Pharmaceutical Preparations
2835	In Vitro and In Vivo Diagnostic Substances
2836	Biological Products, Except Diagnostic Substances
2841	Soap and Other Detergents, Except Specialty Cleaners
2842	Specialty Cleaning, Polishing, and Sanitation Preparations
2843	Surface Active Agents, Finishing Agents, Sulfonated Oils, and Assistants
2844	Perfumes, Cosmetics, and Other Toilet Preparations
2851	Paints, Varnishes, Lacquers, Enamels, and Allied Products
2861	Gum and Wood Chemicals
2865	Cyclic Organic Crudes and Intermediates, and Organic Dyes and Pigments
2869	Industrial Organic Chemicals, Not Elsewhere Classified
2873	Nitrogenous Fertilizers
2874	Phosphatic Fertilizers
2875	Fertilizers, Mixing Only
2879	Pesticides and Agricultural Chemicals, Not Elsewhere Classified
2891	Adhesives and Sealants
2892	Explosives
2893	Printing Ink
2895	Carbon Black
2899	Chemicals and Chemical Preparations, Not Elsewhere Classified

SIC 35 – INDUSTRIAL AND COMMERCIAL MACHINERY AND COMPUTER EQUIPMENT

SIC
Industry
Number   Industry

3511	Steam, Gas, and Hydraulic Turbines, and Turbine Generator Set Units
3519	Internal Combustion Engines, Not Elsewhere Classified
3523	Farm Machinery and Equipment
3524	Lawn and Garden Tractors and Home Lawn and Garden Equipment
3531	Construction Machinery and Equipment
3532	Mining Machinery and Equipment, Except Oil and Gas Field Machinery and Equipment
3533	Oil and Gas Field Machinery and Equipment
3534	Elevators and Moving Stairways
3535	Conveyors and Conveying Equipment
3536	Overhead Traveling Cranes, Hoists, and Monorail Systems
3537	Industrial Trucks, Tractors, Trailers, and Stackers
3541	Machine Tools, Metal Cutting Types
3542	Machine Tools, Metal Forming Types
3543	Industrial Patterns
3544	Special Dies and Tools, Die Sets, Jigs and Fixtures, and Industrial Molds
3545	Cutting Tools, Machine Tool Accessories, and Machinist Precision Measuring Devices
3546	Power-driven Hand Tools
3547	Rolling Mill Machinery and Equipment
3548	Electric and Gas Welding and Soldering Equipment
3549	Metalworking Machinery, Not Elsewhere Classified
3552	Textile Machinery
3553	Woodworking Machinery
3554	Paper Industries Machinery
3555	Printing Trades Machinery and Equipment
3556	Food Products Machinery
3559	Special Industry Machinery, Not Elsewhere Classified
3561	Pumps and Pumping Equipment
3562	Ball and Roller Bearings
3563	Air and Gas Compressors
3564	Industrial and Commercial Fans and Blowers and Air Purification Equipment
3565	Packaging Machinery
3566	Speed Changers, Industrial High-speed Drives, and Gears

3567	Industrial Process Furnaces and Ovens
3568	Mechanical Power Transmission Equipment, Not Elsewhere Classified
3569	General Industrial Machinery and Equipment, Not Elsewhere Classified
3571	Electronic Computers
3572	Computer Storage Devices
3575	Computer Terminals
3577	Computer Peripheral Equipment, Not Elsewhere Classified
3578	Calculating and Accounting Machines, Except Electronic Computers
3579	Office Machines, Not Elsewhere Classified
3581	Automatic Vending Machines
3582	Commercial Laundry, Dry-cleaning, and Pressing Machines
3585	Air-conditioning and Warm Air Heating Equipment and Commercial and Industrial Refrigeration Equipment
3586	Measuring and Dispensing Pumps
3589	Service Industry Machinery, Not Elsewhere Classified
3592	Carburetors, Pistons, Piston Rings, and Valves
3593	Fluid Power Cylinders and Actuators
3594	Fluid Power Pumps and Motors
3596	Scales and Balances, Except Laboratory
3599	Industrial and Commercial Machinery and Equipment, Not Elsewhere Classified

SIC 36 - ELECTRONIC AND OTHER ELECTRICAL EQUIPMENT

SIC *Industry* *Number*	*Industry*
3612	Power, Distribution, and Specialty Transformers
3613	Switchgear and Switchboard Apparatus
3621	Motors and Generators
3624	Carbon and Graphite Products
3625	Relays and Industrial Controls
3629	Electrical Industrial Apparatus, Not Elsewhere Classified
3631	Household Cooking Equipment
3632	Household Refrigerators and Home and Farm Freezers
3633	Household Laundry Equipment
3634	Electric Housewares and Fans

3635	Household Vacuum Cleaners
3639	Household Appliances, Not Elsewhere Classified
3641	Electric Lamp Bulbs and Tubes
3643	Current-carrying Wiring Devices
3644	Noncurrent-carrying Wiring Devices
3645	Residential Electric Lighting Fixtures
3646	Commercial, Industrial, and Institutional Electric Lighting Fixtures
3647	Vehicular Lighting Equipment
3648	Lighting Equipment, Not Elsewhere Classified
3651	Household Audio and Video Equipment
3652	Phonograph Records and Prerecorded Audio Tapes and Disks
3661	Telephone and Telegraph Apparatus
3663	Radio and Television Broadcasting and Communications Equipment
3669	Communications Equipment, Not Elsewhere Classified
3671	Electron Tubes
3672	Printed Circuit Boards
3674	Semiconductors and Related Devices
3675	Electronic Capacitors
3676	Electronic Resistors
3677	Electronic Coils, Transformers, and Other Inductors
3678	Electronic Connectors
3679	Electronic Components, Not Elsewhere Classified
3691	Storage Batteries
3692	Primary Batteries, Dry and Wet
3694	Electrical Equipment for Internal Combustion Engines
3695	Magnetic and Optical Recording Media
3699	Electrical Machinery, Equipment, and Supplies, Not Elsewhere Classified

SIC 38 – INSTRUMENTS AND RELATED PRODUCTS

SIC Industry Number	*Industry*
3812	Search, Detection, Navigation, Guidance, Aeronautical, and Nautical Systems and Instruments
3821	Laboratory Apparatus and Furniture
3822	Automatic Controls for Regulating Residential and Commercial Environments and Appliances

3823	Industrial Instruments for Measurement, Display, and Control of Process Variables; and Related Products
3824	Totalizing Fluid Meters and Counting Devices
3825	Instruments for Measuring and Testing of Electricity and Electrical Signals
3826	Laboratory Analytical Instruments
3827	Optical Instruments and Lenses
3829	Measuring and Controlling Devices, Not Elsewhere Classified
3841	Surgical and Medical Instruments and Apparatus
3842	Orthopedic, Prosthetic, and Surgical Appliances and Supplies
3843	Dental Equipment and Supplies
3844	X-Ray Apparatus and Tubes and Related Irradiation Apparatus
3845	Electromedical and Electrotherapeutic Apparatus
3851	Ophthalmic Goods
3861	Photographic Equipment and Supplies
3873	Watches, Clocks, Clockwork Operated Devices, and Parts

SIC 87 - ENGINEERING, ACCOUNTING, RESEARCH, MANAGEMENT, AND RELATED SERVICES

SIC
Industry
Number   *Industry*

8711	Engineering Services
8712	Architectural Services
8713	Surveying Services
8721	Accounting, Auditing, and Bookkeeping Services
8731	Commercial Physical and Biological Research
8732	Commercial Economic, Sociological, and Educational Research
8733	Noncommercial Research Organizations
8734	Testing Laboratories
8741	Management Services
8742	Management Consulting Services
8743	Public Relations Services
8744	Facilities Support Management Services
8748	Business Consulting Services, Not Elsewhere Classified

APPENDIX B

GEOGRAPHIC DISTRIBUTION OF INNOVATION AND INDICATORS OF THE TECHNOLOGICAL INFRASTRUCTURE IN 1990-95

The following maps show the distributions of innovation and indicators of the technological infrastructure across metropolitan areas in the United States during 1990-95. Innovation in high-technology industries is measured by both SBIR Phase II awards and patents. The technological infrastructure within a metropolitan area is measured by the following indicators: population density, number of R&D labs, business services employment, the concentration of industrial employment by industry, and academic R&D expenditures by industry.

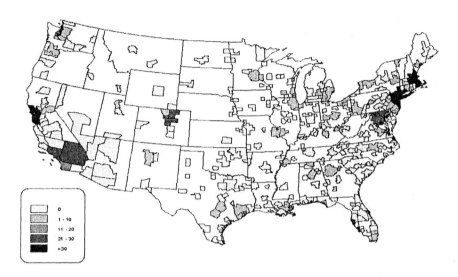

Figure 4. Number of Phase II Awards, 1990-95
SIC 28 – Chemicals and Allied Product

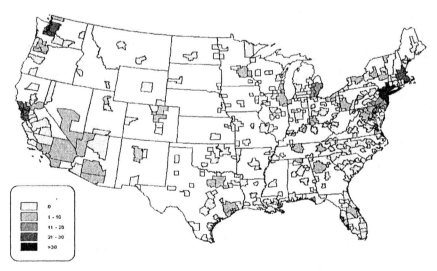

Figure 5. Number of SBIR Phase II Awards, 1990-95
SIC 35 – Industrial Machinery

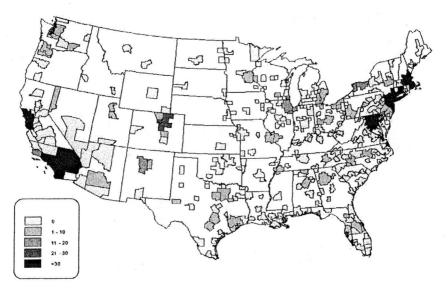

Figure 6. Number of SBIR Phase II Awards, 1990-95
SIC 36 – Electronics and Electrical Equipment

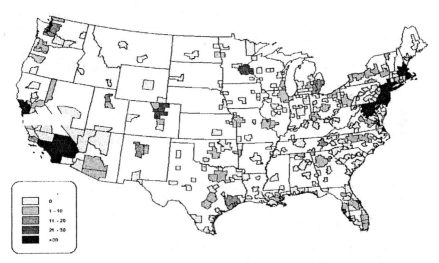

Figure 7. Number of SBIR Phase II Awards, 1990-95
SIC 38 – Scientific Instruments

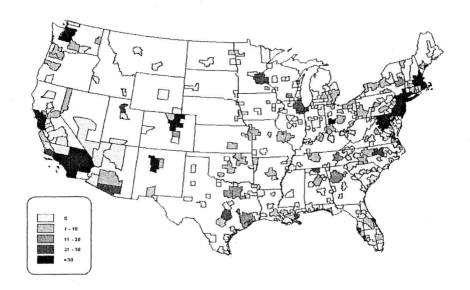

Figure 8. Number of SBIR Phase II Awards, 1990-95
SIC 87 – Research Services

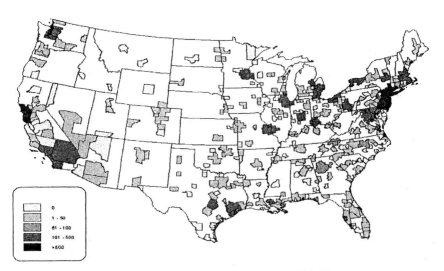

Figure 9. Number of Utility Patents, 1990-95
SIC 28 – Chemicals and Allied Products

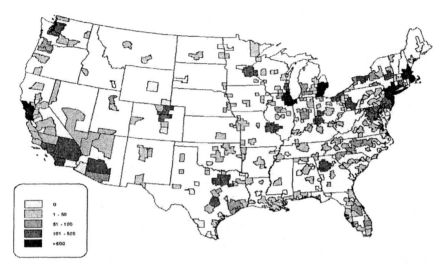

Figure 10. Number of Utility Patents, 1990-95
SIC 35 – Industrial Machinery

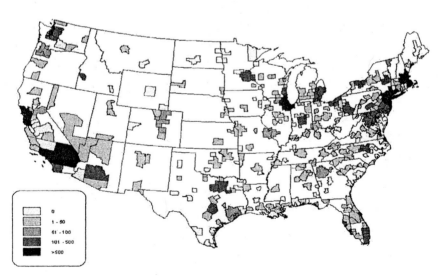

Figure 11. Number of Utility Patents, 1990-95
SIC 36 – Electronics and Electrical Equipment

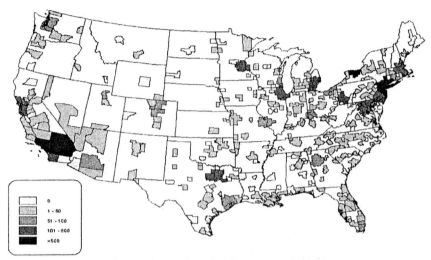

Figure 12. Number of Utility Patents, 1990-95
SIC 38 – Scientific Instruments

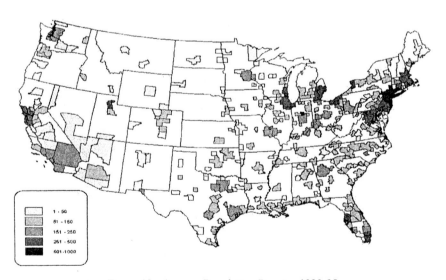

Figure 13. Average Population Density, 1990-95

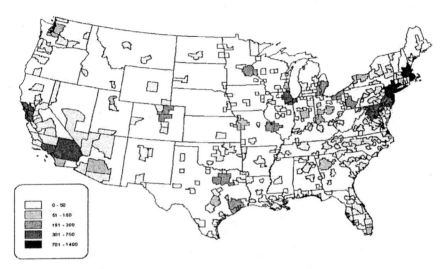

Figure 14. Average Number of R&D Labs, 1990-95

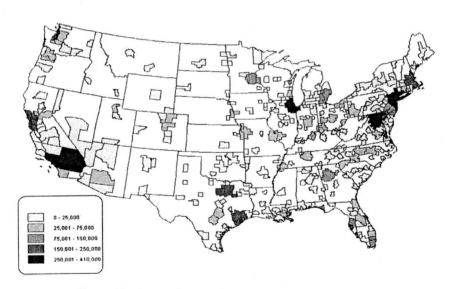

Figure 15. Average Business Services Employment, 1990-95

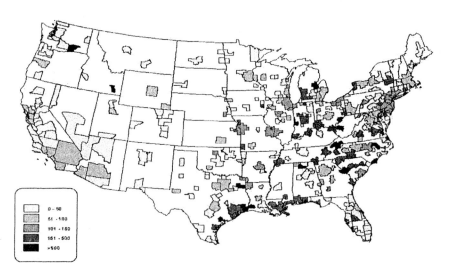

Figure 16. Concentration of Industrial Employment, 1990-95
SIC 28 – Chemicals and Allied Products

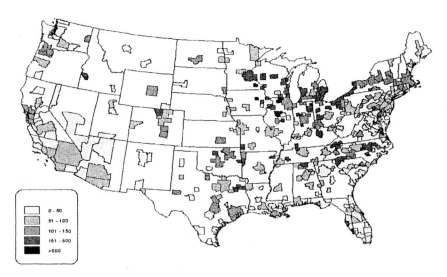

Figure 17. Concentration of Industrial Employment, 1990-95
SIC 35 – Industrial Machinery

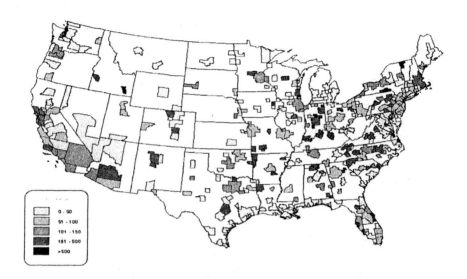

Figure 18. Concentration of Industrial Employment, 1990-95
SIC 36 – Electronics and Electrical Equipment

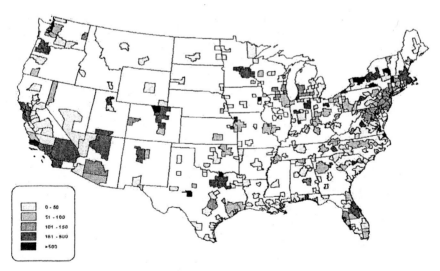

Figure 19. Concentration of Industrial Employment, 1990-95
SIC 38 – Scientific Instruments

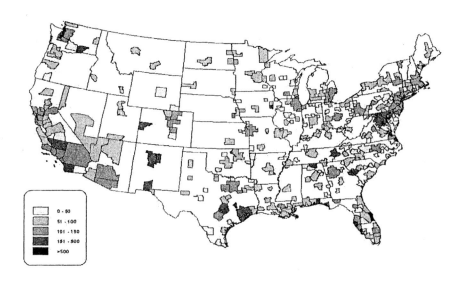

*Figure 20. Concentration of Industrial Employment, 1990-95
SIC 87 – Research Services*

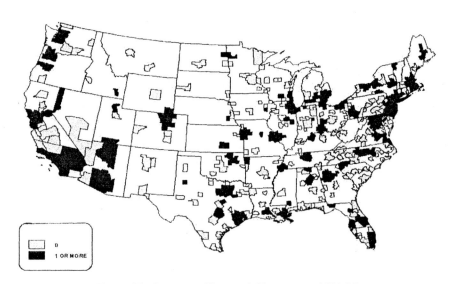

Figure 21. Presence of Research Universities, 1990-95

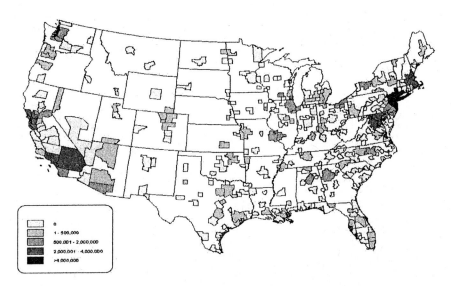

Figure 22. Total Academic R&D Expenditures, 1990-95 (thousands of 1992 dollars)
SIC 28 – Chemicals and Allied Products

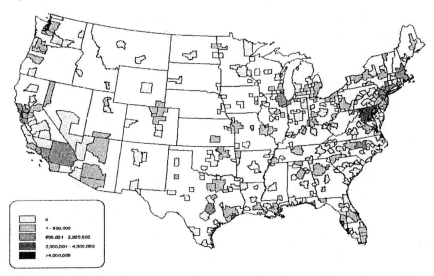

Figure 23. Total Academic R&D Expenditures, 1990-95 (thousands of 1992 dollars)
SIC 35 – Industrial Machinery

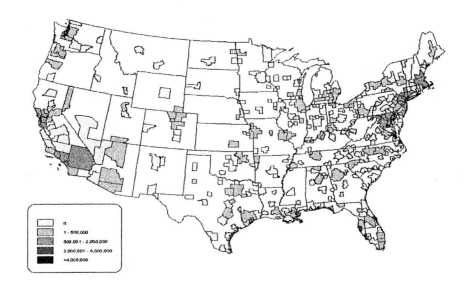

Figure 24. Total Academic R&D Expenditures, 1990-95 (thousands of 1992 dollars)
SIC 36 – Electronics and Electrical Equipment

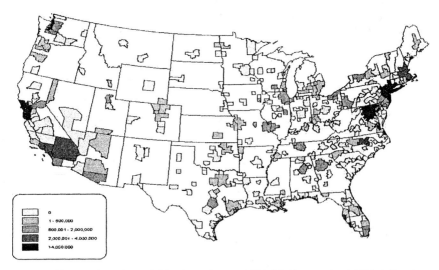

Figure 25. Total Academic R&D Expenditures, 1990-95 (thousands of 1992 dollars)
SIC 38 – Scientific Instruments

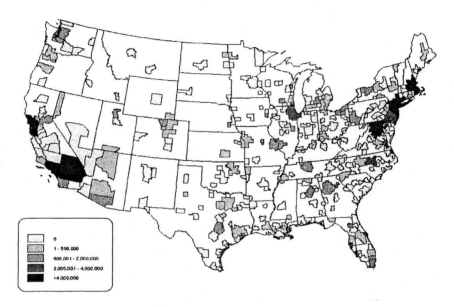

Figure 26. Total Academic R&D Expenditures, 1990-95 (thousands of 1992 dollars)
SIC 87 – Research Services

APPENDIX C

SBIR PHASE II AWARDS BY METROPOLITAN AREA

<u>0 Awards</u>
Abilene, TX
Albany, GA
Alexandria, LA
Altoona, PA
Amarillo, TX
Anniston, AL
Augusta-Aiken, GA-SC
Bangor, ME
Beaumont-Port Arthur, TX
Billings, MT
Bismarck, ND
Boise, ID
Brownsville-Harlingen-San Benito, TX
Canton-Massillon, OH
Casper, WY
Cedar Rapids, IA
Charleston, WV
Cheyenne, WY
Chico-Paradise, CA
Clarksville-Hopkinsville, TN-KY
Columbia, SC

Columbus, GA-AL
Corpus Christi, TX
Cumberland, MD-WV
Danville, VA
Davenport-Moline-Rock Island, IA-IL
Decatur, AL
Decatur, IL
Des Moines, IA
Dothan, AL
Dover, DE
Dubuque, IA
Eau Claire, WI
El Paso, TX
Enid, OK
Erie, PA
Evansville-Henderson, IN-KY
Flagstaff, AZ-UT
Florence, AL
Florence, SC
Fort Myers-Cape Coral, FL
Fort Pierce-Port St. Lucie, FL
Fort Wayne, IN

Gadsden, AL
Glen Falls, NY
Goldsboro, NC
Grand Junction, CO
Grand Rapids-Muskegon-Holland, MI
Great Falls, MT
Green Bay, WI
Greenville-Spartanburg-Anderson, SC
Hattiesburg, MS
Hickory-Morganton-Lenoir, NC
Houma, LA
Jackson, TN
Jacksonville, FL
Jacksonville, NC
Jamestown, NY
Janesville-Beloit, WI
Johnson City-Kingsport-Bristol, TN-
    VA
Johnstown, PA
Jonesboro, AR
Joplin, MO
Kalamazoo-Battle Creek, MI
Killeen-Temple, TX
Kokomo, IN
La Crosse, WI-MN
Lake Charles, LA
Lakeland-Winter Haven, FL
Laredo, TX
Lewiston-Auburn, ME
Lima, OH
Longview-Marshall, TX
Lubbock, TX
Lynchburg, VA
Macon, GA
Mansfield, OH
McAllen-Edinburg-Mission, TX
Medford-Ashland, OR
Merced, CA
Mobile, AL
Modesto, CA
Monroe, LA
Montgomery, AL
Muncie, IN

Myrtle Beach, SC
Naples, FL
Ocala, FL
Owensboro, KY
Panama City, FL
Parkersburg-Marietta, WV-OH
Pensacola, FL
Pittsfield, MA
Peoria-Pekin, IL
Pine Bluff, AR
Pueblo, CO
Punta Gorda, FL
Rapid City, SD
Reading, PA
Redding, CA
Rockford, IL
Rocky Mount, NC
Saint Cloud, MN
Saint Joseph, MO
San Angelo, TX
San Luis Obispo-Atascadero-Pasco
    Robles, CA
Savannah, GA
Scranton-Wilkes/Barre-Hazleton, PA
Sharon, PA
Sheboygan, WI
Sherman-Denison, TX
Shreveport-Bossier City, LA
Sioux City, IA-NE
Sioux Falls, SD
South Bend, IN
Springfield, IL
Springfield, MO
Steubenville-Weirton, OH-WV
Stockton-Lodi, CA
Sumter, SC
Tallahassee, FL
Terre Haute, IN
Texarkana, TX-AR
Tuscaloosa, AL
Tyler, TX
Victoria, TX
Visalia-Tulare-Porterville, CA

Waco, TX
Wausau, WI
Wheeling, WV-OH
Wichita Falls, TX
Williamsport, PA
Wilmington, NC
Youngstown-Warren, OH
Yuma, AZ

1 Award
Anchorage, AK
Appleton-Oshkosh-Neenah, WI
Asheville, NC
Benton Harbor, MI
Biloxi-Gulfport-Pascagoula, MS
Bloomington-Normal, IL
Charleston-North Charleston, SC
Daytona Beach, FL
Fargo-Moorhead, ND-MN
Fayetteville, NC
Fresno, CA
Fort Smith, AR-OK
Greenville, NC
Harrisburg-Lebanon-Carlisle, PA
Huntington-Ashland, WV-KY-OH
Jackson, MI
Lafayette, LA
Odessa-Midland, TX
Pocatello, ID
Rochester, MN
Saginaw-Bay City-Midland, MI
Spokane, WA
Topeka, KS
Tulsa, OK
Waterloo-Cedar Falls, IA
Yakima, WA
York, PA
Yuba City, CA

2 Awards
Bakersfield, CA
Baton Rouge, LA
Bellingham, WA

Charlotte-Gastonia-Rock Hill, NC-SC
Columbia, MO
Duluth-Superior, MN-WI
Elkhart-Goshen, IN
Elmira, NY
Fayetteville-Springdale-Rogers, AR
Grand Forks, ND-MN
Lafayette, IN
Little Rock-North Little Rock, AR
Memphis, TN-AR-MS
Omaha, NE-IA
Roanoke, VA
Sarasota-Bradenton, FL
Wichita, KS

3 Awards
Fort Walton Beach, FL
Iowa City, IA
Kansas City, MO-KS
Lawton, OK
Salinas, CA

4 Awards
Athens, GA
Indianapolis, IN
Lexington, KY
Louisville, KY-IN
New London-Norwich, CT
Providence-Warwick-Pawtucket, RI

5 Awards
Binghamton, NY
Greensboro-Winston Salem-High Point, NC
Miami-Fort Lauderdale, FL
State College, PA

6 Awards
New Orleans, LA
Richmond-Petersburg, VA
Syracuse, NY

7 Awards

Birmingham, AL
Bloomington, IN
Oklahoma City, OK
Provo-Orem, UT
Tampa-St. Petersburg-Clearwater, FL

8 Awards
Cincinnati-Hamilton, OH-KY-IN
Las Vegas, NV-AZ
Richland-Kennewick-Pasco, WA

9 Awards
Lawrence, KS
Nashville, TN
Portland, ME
Reno, NV
St. Louis, MO-IL

10 Awards
Lansing-East Lansing, MI
Las Cruces, NM
Lincoln, NE
Milwaukee-Racine, WI
West Palm Beach-Boca Raton, FL

11 Awards
Champaign-Urbana, IL
Eugene-Springfield, OR
Utica-Rome, NY

13 Awards
Barnstable-Yarmouth, MA
Chattanooga, TN-GA
Columbus, OH
Toledo, OH

14 Awards
Burlington, VT
Charlottesville, VA

16 Awards
Fort Collins-Loveland, CO
Gainesville, FL

Honolulu, HI
Santa Fe, NM
Springfield, MA

18 Awards
Bryan-College Station, TX
Lancaster, PA

19 Awards
Allentown-Bethlehem-Easton, PA

20 Awards
San Antonio, TX

21 Awards
Sacramento-Yolo, CA

24 Awards
Madison, WI

26 Awards
Phoenix-Mesa, AZ
Portland-Salem, OR-WA
Rochester, NY

28 Awards
Norfolk-Virginia Beach-Newport
News, VA-NC

34 Awards
Knoxville, TN
Orlando, FL

36 Awards
Dallas-Fort Worth, TX
Melbourne-Titusville-Palm Bay, FL

37 Awards
Albany-Schenectady, NY

38 Awards
Atlanta, GA
Cleveland-Akron, OH

Colorado Springs, CO

40 Awards
Buffalo-Niagara Falls, NY
Pittsburgh, PA

44 Awards
Tucson, AZ

45 Awards
Austin-San Marcos, TX

57 Awards
Hartford, CT

58 Awards
Raleigh-Durham-Chapel Hill, NC

61 Awards
Houston-Galveston-Brazoria, TX

64 Awards
Santa Barbara-Santa Maria-Lompoc,
CA

68 Awards
Salt Lake City-Ogden, UT

76 Awards
Chicago-Gary-Kenosha, IL-IN-WI
Dayton-Springfield, OH
Minneapolis-St. Paul, MN-WI

77 Awards
Huntsville, AL

79 Awards
Detroit-Ann Arbor-Flint, MI

84 Awards
Albuquerque, NM

112 Awards
Seattle-Tacoma-Bremerton, WA

139 Awards
Philadelphia-Wilmington-Atlantic City,
PA-NJ-DE-MD

154 Awards
Denver-Boulder-Greeley, CO

202 Awards
San Diego, CA

421 Awards
New York-Northern New Jersey-Long
Island, NY-NJ-CT-PA

498 Awards
San Francisco-Oakland-San Jose, CA

507 Awards
Washington-Baltimore, DC-MD-VA-
WV

509 Awards
Los Angeles-Riverside-Orange, CA

882 Awards
Boston-Worcester-Lawrence-Lowell-
Brockton, MA-NH

APPENDIX D

SUPPLEMENTAL EVIDENCE OF FUNDING AGENCY EFFECTS

This appendix presents the estimated results from the negative binomial equation of the hurdle model and from different specifications of the probit equation presented in Chapter 6. Brief discussion of these results and comparison to the base model's findings are in Chapter 6. Table 32 shows the negative binomial results by agency. Tables 33 and 34 present the empirical findings for the hurdle model by agency for all five high-tech industries combined. While this specification loses the variation across industries, it allows the negative binomial equation for the number of Phase II awards by agency to be estimated. Tables 35 through 39 show the empirical findings for the probit model in which UNIVDUM (the variable indicating the presence of research universities) is replaced by UNIVR&D (the variable for the level of university R&D expenditures). Tables 40 through 45 present the empirical results for the probit model that includes an additional variable measuring proximity to military installations. Two specifications are estimated: one adding a variable for the number of military installations located in a given metropolitan area (MSA MILITARY) and the second adding a variable for the number of military installations located in the state(s) in which a given metropolitan area is located (STATE MILITARY).

Table 32. Rate of Phase II Awards by SBIR Funding Agency and Industry

Estimated Coefficients
(Standard Errors)

Variable	SIC 28 DOD	SIC 28 HHS	SIC 35 DOD	SIC 36 DOD	SIC 36 NASA	SIC 38 DOD	SIC 38 HHS	SIC 38 NASA	SIC 87 DOD	SIC 87 HHS	SIC 87 NASA
Constant	0.6057	2.5156	1.1159	2.1482	1.4760	1.1643	1.4594	1.9047	2.7831	1.7819	1.3001
	(0.6604)	(78.1620)	(5.9533)	(0.8400)	(9.3307)	(0.5775)	(0.5433)	(4.6512)	(2.1409)	(1.1666)	(1.4185)
POPDEN	0.1572	-0.2457	0.5472	-0.3250*	-0.3980	-0.1726	-0.03473	-0.3396	-0.1383	-0.3841	-0.2867
	(0.2469)	(10.6526)	(0.5494)	(0.1940)	(0.4259)	(0.2111)	(0.1309)	(0.2810)	(0.2125)	(0.2546)	(0.2089)
R&DLABS	0.07179	0.1205	-0.2324	0.3904	0.2394	0.1268	0.2185	0.2066	0.1936	0.1881	0.2364
	(0.1737)	(16.5950)	(0.7443)	(0.3172)	(1.1869)	(0.1924)	(0.1369)	(0.3120)	(0.4398)	(0.2366)	(0.2565)
EMPSIC73	-0.02305	-0.02474	0.02932	-0.05929	-0.001862	0.03249	-0.06357	-0.01573	-0.03697	-0.005294	-0.01068
	(0.03386)	(1.4039)	(0.1236)	(0.06479)	(0.6930)	(0.04469)	(0.04113)	(0.1735)	(0.1319)	(0.03440)	(0.07947)
UNIVR&D$_i$	0.04750*	0.05498	0.003496	0.1385*	0.01290	0.01700	0.04988**	0.02245	0.03708	0.03949*	0.02688*
	(0.02834)	(0.2997)	(0.09383)	(0.4218)	(0.3148)	(0.02601)	(0.01990)	(0.03300)	(0.04366)	(0.02354)	(0.01220)
EMPCON$_i$	-0.8008**	-0.2185	-0.3820	0.1646	0.5247***	0.2684**	-0.08494	0.04113	-0.03365	-0.06389	0.1492
	(0.3258)	(15.3142)	(0.4714)	(0.1675)	(0.1559)	(0.09263)	(0.1317)	(0.5947)	(0.6778)	(0.4729)	(0.5368)
INVERSE MILLS RATIO	0.1078	-1.02713	-0.1887	-0.5703	-0.5269	-0.3009	-0.2953	-0.6135	-1.1354	-0.4805	-0.3491
	(0.3216)	(55.2198)	(2.8119)	(0.4218)	(4.9982)	(0.3021)	(0.3944)	(2.4660)	(1.1938)	(0.6282)	(0.7144)
α	0.01333	0.01669	0.2741	0.5542**	0.02637	0.3696**	0.3303	0.1264	0.5464***	0.3106	0.1353
	(0.1916)	(0.3920)	(0.2884)	(0.2179)	(1.3663)	(0.1510)	(0.2288)	(0.3379)	(0.1724)	(0.2261)	(0.2206)
Ln L	-42.7	-78.9	-62.0	-128.7	-59.1	-133.3	-103.3	-67.0	-191.6	-110.9	-96.4

*Significant at the 10 percent level
**Significant at the 5 percent level
***Significant at the 1 percent level

Table 33. Likelihood of Phase II Awards by SBIR Funding Agency
High Technology Industries

Variable	Estimated Coefficients (Standard Errors)		
	DOD	HHS	NASA
Constant	-1.06881	-1.5424	-1.3742
	(0.13531)	(0.1586)	(0.1513)
POPDEN	0.08116	0.1326	0.04184
	(0.1043)	(0.1075)	(0.1147)
R&DLABS	2.4474***	0.9677*	1.9149***
	(0.6920)	(0.5309)	(0.5852)
EMPSIC73	-0.1199	0.02303	-0.05150
	(0.089775)	(0.08070)	(0.08305)
UNIVDUM$_i$	0.7032***	1.1768***	0.7314***
	(0.2107)	(0.2138)	(0.2172)
Ln L	-128.0	-101.4	-106.6
χ^2	94.3	110.2	97.7

*Significant at the 10 percent level
**Significant at the 5 percent level
***Significant at the 1 percent level

Table 34. Rate of Phase II Awards by SBIR Funding Agency
High Technology Industries

Variable	Estimated Coefficients (Standard Errors)		
	DOD	HHS	NASA
Constant	3.5461	2.06883	2.2920
	(0.4419)	(0.3392)	(0.4190)
POPDEN	-0.5013***	-0.2462*	-0.4080**
	(0.1928)	(0.1365)	(0.2058)
R&DLABS	0.4142	0.1994	0.2773
	(0.4558)	(0.1649)	(0.2879)
EMPSIC73	-0.02878	0.008455	-0.01326
	(0.1507)	(0.03346)	(0.09738)
UNIVR&D$_i$	0.04525	0.05985**	0.05382*
	(0.04884)	(0.02921)	(0.03055)
INVERSE	-1.3023***	-0.5556**	-0.7099*
MILLS	(0.36001)	(0.2803)	(0.4097)
RATIO			
α	0.6990***	0.4947***	0.4311***
	(0.1645)	(0.1601)	(0.1424)
Ln L	-304.0	-198.8	-180.6

*Significant at the 10 percent level
**Significant at the 5 percent level
***Significant at the 1 percent level

Table 35. Likelihood of Phase II Awards Using University R&D Expenditures
SIC 28 – Chemicals and Allied Products

Variable	Estimated Coefficients (Standard Errors)		
	DOD	HHS	NASA
Constant	-1.9641	-1.6553	-1.7483
	(0.2664)	(0.1832)	(0.3567)
POPDEN	0.01906	0.2406**	-0.7808
	(0.1733)	(0.1002)	(0.4872)
R&DLABS	1.8880***	0.2304	1.06208
	(0.6083)	(0.3843)	(0.7500)
EMPSIC73	-0.1714**	0.03281	-0.05274
	(0.09838)	(0.06512)	(0.1176)
UNIVR&D$_i$	0.1170***	0.1548***	0.005595
	(0.04356)	(0.04947)	(0.06874)
EMPCON$_i$	-0.1387	-0.09218	-0.05663
	(0.1789)	(0.09521)	(0.1532)
Ln L	-41.9	-77.9	-20.5
χ^2	74.0	68.2	31.3

*Significant at the 10 percent level
**Significant at the 5 percent level
***Significant at the 1 percent level

Table 36. Likelihood of Phase II Awards Using University R&D Expenditures
SIC 35 – Industrial Machinery

Variable	Estimated Coefficients (Standard Errors)		
	DOD	HHS	NASA
Constant	-1.7278	-2.6273	-1.9013
	(0.2226)	(0.5718)	(0.2506)
POPDEN	0.02695	-0.03361	0.1435
	(0.1430)	(0.3911)	(0.1199)
R&DLABS	0.4068	1.1517**	0.2211
	(0.3629)	(0.5440)	(0.3435)
EMPSIC73	0.03488	-0.009888	0.03212
	(0.06671)	(0.07919)	(0.06481)
UNIVR&D$_i$	0.05617	-0.2299	0.09801
	(0.1115)	(0.1426)	(0.1072)
EMPCON$_i$	0.04039	-0.2008	-0.04858
	(0.1178)	(0.4914)	(0.1073)
Ln L	-65.9	-10.8	-53.2
χ^2	44.3	43.6	41.6

*Significant at the 10 percent level
**Significant at the 5 percent level
***Significant at the 1 percent level

Table 37. Likelihood of Phase II Awards Using University R&D Expenditures
SIC 36 – Electronics and Electrical Equipment

Variable	Estimated Coefficients (Standard Errors)		
	DOD	HHS	NASA
Constant	-1.5660	-1.8820	-1.7532
	(0.1675)	(0.2110)	(0.2078)
POPDEN	0.03451	-0.06008	0.05730
	(0.1287)	(0.1769)	(0.1585)
R&DLABS	0.8424*	0.3746	0.7011*
	(0.4716)	(0.3934)	(0.3896)
EMPSIC73	0.07058	-0.008438	0.04614
	(0.07543)	(0.07651)	(0.06386)
UNIVR&D$_i$	0.5428***	0.2002	0.2144
	(0.1649)	(0.1647)	(0.1659)
EMPCON$_i$	0.007559	0.06384	-0.06117
	(0.07842)	(0.08754)	(0.1108)
Ln L	-85.5	-46.8	-64.1
χ^2	82.9	33.8	60.8

*Significant at the 10 percent level
**Significant at the 5 percent level
***Significant at the 1 percent level

The Geography of Small Firm Innovation

Table 38. Likelihood of Phase II Awards Using University R&D Expenditures
SIC 38 – Scientific Instruments

Variable	Estimated Coefficients (Standard Errors)		
	DOD	HHS	NASA
Constant	-1.3892	-1.6761	-1.7419
	(0.1749)	(0.1952)	(0.1861)
POPDEN	-0.1534	-0.08779	0.06799
	(0.1651)	(0.1657)	(0.1240)
R&DLABS	1.2559**	0.7331	0.4191
	(0.5253)	(0.4861)	(0.3479)
EMPSIC73	0.05311	0.1030	0.05788
	(0.08042)	(0.07887)	(0.06411)
UNIVR&D$_i$	0.1039**	0.1489***	-0.001942
	(0.04091)	(0.04255)	(0.03545)
EMPCON$_i$	0.08245	0.07880	0.1227*
	(0.06353)	(0.07648)	(0.06892)
Ln L	-93.2	-73.9	-73.8
χ^2	85.1	93.4	53.7

Significant at the 10 percent level
*Significant at the 5 percent level
**Significant at the 1 percent level

*Table 39. Likelihood of Phase II Awards Using University R&D Expenditures
SIC 87, Research Services*

| | Estimated Coefficients (Standard Errors) | | |
Variable	DOD	HHS	NASA
Constant	-1.9529	-2.2873	-2.3934
	(0.2259)	(0.2648)	(0.2657)
POPDEN	0.01749	0.04823	0.1353
	(0.1260)	(0.1354)	(0.1187)
R&DLABS	0.6445*	0.6428	0.9277*
	(0.3775)	(0.4789)	(0.4790)
EMPSIC73	0.03098	0.08270)	-0.06169
	(0.06566)		(0.07435)
UNIVR&D$_i$	0.05502*	0.1265***	0.1133***
	(0.02935)	(0.03497)	(0.03356)
EMPCON$_i$	0.9988***	0.6976***	0.8473***
	(0.1922)	(0.1916)	(0.1988)
Ln L	-103.4	-70.5	-71.3
χ^2	95.2	112.9	98.6

*Significant at the 10 percent level
**Significant at the 5 percent level
***Significant at the 1 percent level

Table 40. Likelihood of Phase II Awards Including Military Installations
by SBIR Funding Agency, High Technology Industries

	Estimated Coefficients (Standard Errors)					
Variable	DOD		HHS		NASA	
Constant	-1.09759	-1.03679	-1.5818	-1.4227	-1.3939	-1.4171
	(0.1382)	(0.1728)	(0.1642)	(0.1938)	(0.1536)	(0.1935)
POPDEN	0.03919	0.08098	0.09933	0.1346	-0.00004286	0.04195
	(0.1133)	(0.1043)	(0.1172)	(0.1073)	(0.1265)	(0.1149)
R&DLABS	2.4877***	2.4302***	0.9826**	0.9187*	1.9333***	1.9392***
	(0.6976)	(0.6937)	(0.5303)	(0.5304)	(0.5891)	(0.5908)
EMPSIC73	-0.1355	-0.1161	0.005610	0.03582	-0.06332	-0.05658
	(0.09086)	(0.09063)	(0.08116)	(0.08166)	(0.08394)	(0.08434)
UNIVDUM	0.6773***	0.7017***	1.1516***	1.1845***	0.7050***	0.7328***
	(0.2123)	(0.2107)	(0.2151)	(0.2146)	(0.2189)	(0.2173)
MSA	0.08145		0.08351		0.06940	
MILITARY	(0.05645)		(0.05469)		(0.05601)	
STATE		-0.002045		-0.008465		0.002675
MILITARY		(0.006919)		(0.008219)		(0.007411)
Ln L	-126.9	-127.9	-100.2	-100.8	-105.8	-106.5
χ^2	96.4	94.4	112.6	111.3	99.3	97.8

*Significant at the 10 percent level
**Significant at the 5 percent level
***Significant at the 1 percent level

Table 41. Likelihood of Phase II Awards Including Military Installations
by SBIR Funding Agency, SIC 28 – Chemicals and Allied Products

Variable	Estimated Coefficients (Standard Errors)					
	DOD		HHS		NASA	
Constant	-2.1086	-1.9233	-1.8628	-1.5949	-1.9051	-2.1653
	(0.2983)	(0.3380)	(0.2104)	(0.2440)	(0.4688)	(0.5165)
POPDEN	-0.06426	0.009089	0.2058	0.2291**	-1.1222*	-0.9144*
	(0.2039)	(0.1719)	(0.1066)	(0.1011)	(0.6428)	(0.4931)
R&DLABS	1.8251***	1.7425***	0.3478	0.3387	1.3425	1.1015*
	(0.6014)	(0.5794)	(0.3353)	(0.3183)	(0.8341)	(0.6432)
EMPSIC73	-0.1744*	-0.1436	0.003580	0.03942	-0.04494	-0.07446
	(0.09508)	(0.09120)	(0.06271)	(0.06130)	(0.1295)	(0.1104)
UNIVDUM	0.6390**	0.7015**	0.7574	0.7978***	0.8322*	0.8563*
	(0.3219)	(0.3209)	(0.2424)	(0.2417)	(0.4886)	(0.4952)
EMPCON	-0.1288	-0.1600	-0.07399	-0.08930	-0.05650	-0.04181
	(0.1827)	(0.1960)	(0.09391)	(0.09980)	(0.1792)	(0.1657)
MSA MILITARY	0.07783 (0.06601)		0.06876 (0.05138)		-0.07792 (0.07597)	
STATE MILITARY		-0.01272 (0.01318)		-0.01745 (0.01062)		0.007302 (0.01383)
Ln L	-42.2	-42.4	-77.8	-77.1	-18.3	-18.8
χ^2	73.4	73.0	68.4	69.7	35.6	34.6

*Significant at the 10 percent level
**Significant at the 5 percent level
***Significant at the 1 percent level

Table 42. Likelihood of Phase II Awards Including Military Installations
by SBIR Funding Agency, SIC 35 – Industrial Machinery

	Estimated Coefficients (Standard Errors)				
Variable	DOD		HHS[1]	NASA	
Constant	-1.8652	-1.7997		-2.03234	-1.9181
	(0.2433)	(0.2974)		(0.2764)	(0.3422)
POPDEN	-0.08294	-0.005447		0.08682	0.1072
	(0.1829)	(0.1540)		(0.1347)	(0.1265)
R&DLABS	0.4837	0.4833		0.3295	0.3611
	(0.3555)	(0.3328)		(0.1347)	(0.2932)
EMPSIC73	0.009875	0.02040		0.01208	0.02217
	(0.06971)	(0.06641)		(0.06161)	(0.2932)
UNIVDUM	0.2625	0.3350		0.3261	0.3726
	(0.2875)	(0.2756)		(0.3015)	(0.2933)
EMPCON	0.09306	0.04846		-0.01625	-0.06783
	(0.1198)	(0.1234)		(0.1603)	(0.1712)
MSA MILITARY	0.09908*			0.05450	
	(0.05637)			(0.04861)	
STATE MILITARY		0.0001607			-0.002936
		(0.009408)			(0.01076)
Ln L	-63.7	-65.4		-52.3	-52.9
χ^2	48.7	45.4		43.5	42.2

[1]Model inestimable due to perfect correlation between UNIVDUM and PH2DUM
*Significant at the 10 percent level
**Significant at the 5 percent level
***Significant at the 1 percent level

Table 43. Likelihood of Phase II Awards Including Military Installations
by SBIR Funding Agency, SIC 36 – Electronics and Electrical Equipment

Variable	\multicolumn{2}{c}{Estimated Coefficients (Standard Errors)}				
	DOD		HHS		NASA	
Constant	-1.7112	-1.6239	-2.0028	-1.8435	-1.8318	-1.6894
	(0.1917)	(0.2271)	(0.2530)	(0.3060)	(0.2296)	(0.2699)
POPDEN	0.003792	0.01328	-0.1500	-0.1512	0.001296	0.02470
	(0.1396)	(0.1339)	(0.1855)	(0.1932)	(0.1732)	(0.1671)
R&DLABS	0.8685*	0.8390*	0.5238	0.5506	0.7246*	0.6939*
	(0.4610)	(0.4614)	(0.3304)	(0.3488)	(0.3839)	(0.3717)
EMPSIC73	0.04248	0.05161	-0.02305	-0.001904	0.02427	0.03826
	(0.07521)	(0.07557)	(0.6778)	(0.06818)	(0.06532)	(0.06518)
UNIVDUM	0.8453***	0.8547***	0.4159	0.4290	0.5027*	0.5222*
	(0.2356)	(0.2344)	(0.3068)	(0.3060)	(0.2807)	(0.2782)
EMPCON	0.0009248	-0.003822	0.08364	0.07359	-0.07309	-0.08315
	(0.07924)	(0.07771)	(0.08711)	(0.08770)	(0.1207)	(0.1196)
MSA MILITARY	0.01040 (0.04311)		0.04293 (0.05121)		0.02753 (0.05690)	
STATE MILITARY		-0.005587 (0.008745)		-0.008980 (0.01263)		-0.009195 (0.01065)
Ln L	-84.5	-84.3	-46.4	-46.5	-63.1	-62.8
χ^2	85.0	85.4	34.4	34.3	62.9	63.5

*Significant at the 10 percent level
**Significant at the 5 percent level
***Significant at the 1 percent level

Table 44. Likelihood of Phase II Awards Including Military Installations
by SBIR Funding Agency, SIC 38 – Scientific Instruments

Variable	Estimated Coefficients (Standard Errors)					
	DOD		HHS		NASA	
Constant	-1.5609	-1.6213	-1.9437	-1.8794	-1.9147	-1.8916
	(0.1857)	(0.2226)	(0.2225)	(0.2516)	(0.2114)	(0.2455)
POPDEN	-0.3083*	-0.2003	-0.1949	-0.1267	-0.03166	0.02659
	(0.16998)	(0.1684)	(0.1741)	(0.1683)	(0.1549)	(0.1369)
R&DLABS	1.2616**	1.2432**	0.7429	0.7263	0.4622	0.4306
	(0.5230)	(0.5222)	(0.4673)	(0.4744)	(0.3533)	(0.3267)
EMPSIC73	0.01929	0.02993	0.07446	0.08643	0.01783	0.03099
	(0.08208)	(0.08153)	(0.07809)	(0.07891)	(0.06779)	(0.06448)
UNIVDUM	0.7970***	0.8339***	1.1502***	1.1592***	0.5664**	0.6136**
	(0.2332)	(0.2310)	(0.2559)	(0.2537)	(0.2594)	(0.2544)
EMPCON	0.08577	0.07438	0.07289	0.07002	0.1029	0.1003
	(0.06608)	(0.06602)	(0.07612)	(0.07626)	(0.07172)	(0.07101)
MSA MILITARY	0.1343**		0.07118		0.08416	
	(0.06172)		(0.05739)		(0.05322)	
STATE MILITARY		0.006726		-0.002915		0.0004388
		(0.007425)		(0.009151)		(0.008810)
Ln L	-87.9	-90.1	-69.5	-70.2	-69.6	-70.9
χ^2	95.7	91.3	102.1	100.7	62.0	59.4

*Significant at the 10 percent level
**Significant at the 5 percent level
***Significant at the 1 percent level

Table 45. Likelihood of Phase II Awards Including Military Installations by SBIR Funding Agency, SIC 87 – Research Services

	Estimated Coefficients (Standard Errors)					
Variable	DOD		HHS		NASA	
Constant	-2.03065	-1.9933	-2.4011	-2.2912	-2.6307	-2.5880
	(0.2363)	(0.2562)	(0.2852)	(0.3056)	(0.3078)	(0.3274)
POPDEN	-0.07721	0.0016998	-0.004281	0.03265	-0.01772	0.1254
	(0.1510)	(0.1301)	(0.1547)	(0.1394)	(0.1673)	(0.1229)
R&DLABS	0.73664*	0.6366*	0.6561	0.5931	1.0604**	0.9004**
	(0.3909)	(0.3756)	(0.4564)	(0.45503)	(0.4581)	(0.4501)
EMPSIC73	-	0.02338	0.07007	0.08809	-0.1110	-0.07065
	0.0007319	(0.06694)	(0.07505)	(0.07509)	(0.07713)	(0.07238)
	(0.06891)					
UNIVDUM	0.4839**	0.5074**	0.8202***	0.8357***	0.9242***	0.9571***
	(0.2252)	(0.2231)	(0.24998)	(0.2499)	(0.2588)	(0.2521)
EMPCON	0.9883***	1.03483***	0.7275***	0.7641***	0.8731***	0.9348***
	(0.1939)	(0.1943)	(0.2049)	(0.2047)	(0.2098)	(0.2093)
MSA MILITARY	0.09470*		0.03754		0.1537**	
	(0.05747)		(0.05475)		(0.06403)	
STATE MILITARY		-0.003309		-0.009524		-0.003549
		(0.007369)		(0.009557)		(0.009102)
Ln L	-101.3	-102.7	-73.1	-72.9	-67.9	-71.2
χ^2	99.3	96.6	107.6	108.2	105.3	98.7

*Significant at the 10 percent level
**Significant at the 5 percent level
***Significant at the 1 percent level

APPENDIX E

LINKING STANDARD INDUSTRIAL CLASSIFICATIONS TO PATENT CLASSIFICATIONS

The following 41 SIC codes are linked to patent classifications by the U.S. Patent and Trademark Office. Patents not classified as belonging to these 41 SIC codes are included in the catch-all category, "All Other SICs." For further information, refer to the *Concordance Between the Standard Industrial Code (SIC) Classification System and the U.S. Patent Classification (USPC) System* (USPTO 2000).

SIC CODE	PRODUCT FIELD
1329	Petroleum and Natural Gas Extraction and Refining
20	Food and Kindred Products
22	Textile Mill Products
281	Industrial Inorganic Chemistry
282	Plastics Materials and Synthetic Resins
283	Drugs and Medicines
284	Soaps, Detergents, Cleaners, Perfumes, Cosmetics and Toiletries
285	Paints, Varnishes, Lacquers, Enamels and Allied Products
286	Industrial Organic Chemistry
287	Agricultural Chemicals
289	Miscellaneous Chemical Products
30	Rubber and Miscellaneous Plastics Products
32	Stone, Clay, Glass and Concrete Products

331+	Primary Ferrous Products (331, 332, 3399, 3462)
333+	Primary and Secondary Non-ferrous Metals (333-336, 339, 3463, excluding 3399)
34-	Fabricated Metal Products (excluding 3462, 3463, 348)
348+	Ordinance Except Missiles (348, 3795)
351	Engines and Turbines
352	Farm and Garden Machinery and Equipment
353	Construction, Mining, and Material Handling Machinery and Equipment
354	Metal Working Machinery and Equipment
355	Special Industry Machinery, Except Metal Working
356	General Industry Machinery and Equipment
357	Office Computing and Accounting Machines
358	Refrigeration and Service Industry Machinery
359	Miscellaneous Machinery, Except Electrical
361+	Electrical Transmission and Distribution Equipment (361, 3825)
362	Electrical Industrial Apparatus
363	Household Appliances
364	Electrical Lighting and Wiring Equipment
365	Radio and Television Receiving Equipment, Except Communication Types
366+	Electronic Components and Accessories and Communications Equipment (366-367)
369	Miscellaneous Electrical Machinery, Equipment and Supplies
371	Motor Vehicles and Other Motor Vehicle Equipment
372	Aircraft and Parts
373	Ship and Boat Building and Repairing
374	Railroad Equipment
375	Motorcycles, Bicycles, and Parts
376	Guided Missiles and Space Vehicles and Parts
379-	Miscellaneous Transportation Equipment (379, excluding 3795)
38-	Professional and Scientific Instruments (excluding 3825)

APPENDIX F

UTILITY PATENTS BY METROPOLITAN AREA

0 Patents
Abilene, TX
Alexandria, LA
Anniston, AL
Brownsville-Harlingen-San Benito,
   TX
Casper, WY
Cumberland, MD-WV
Dothan, AL
Jackson, TN
Johnstown, PA
Laredo, TX
Lawton, OK
Merced, CA
Pocatello, ID
St. Cloud, MN
Texarkana, TX-AR
Yuma, AZ

1 Patent
Albany, GA
Bangor, ME
Bismarck, ND
Danville, VA

Decatur, AL
Fort Walton Beach, FL
Gadsden, AL
Goldsboro, NC
McAllen-Edinburg-Mission, TX
Owensboro, KY
Pine Bluff, AR
San Angelo, TX
Sumter, SC
Tuscaloosa, AL
Yuba City, CA

2 Patents
Billings, MT
Cheyenne, WY
Columbus, GA-AL
Enid, OK
Grand Forks, ND-MN
Grand Junction, CO
Great Falls, MT
Jacksonville, NC
Jonesboro, AR
Pueblo, CO
Rapid City, SD

Savannah, GA
Sioux City, IA-NE

3 Patents
Corpus Christi, TX
Killeen-Temple, TX
Lewiston-Auburn, ME
Redding, CA
St. Joseph, MO
Waco, TX

4 Patents
Clarksville-Hopkinsville, TN-KY
Dover, DE
Jamestown, NY
Monroe, LA
Myrtle Beach, SC
Odessa-Midland, TX
Punta Gorda, FL
Rocky Mount, NC
Sharon, PA

5 Patents
Bellingham, WA
Biloxi-Gulfport-Pascagoula, MS
Flagstaff, AZ-UT
Florence, AL
Huntington-Ashland, WV-KY-OH
Jackson, MS
Tallahassee, FL
Tyler, TX
Victoria, TX
Wheeling, WV-OH
Yakima, WA

6 Patents
Amarillo, TX
Fayetteville, NC
Fort Smith, AR-OK
Hattiesburg, MS
Houma, LA
La Crosse, WI-MN

Panama City, FL
Sioux Falls, SD
Topeka, KS
Visalia-Tulare-Porterville, CA
Wichita Falls, TX

7 Patents
Altoona, PA
Anchorage, AK
Columbia, MO
Decatur, IL
Dubuque, IA
Joplin, MO
Steubenville-Weirton, OH-WV

8 Patents
Chico-Paradise, CA
Florence, SC
Las Cruces, NM
Medford-Ashland, OR
Ocala, FL

9 Patents
Duluth-Superior, MN-WI
Greenville, NC
Naples, FL

10 Patents
Bloomington-Normal, IL
Fargo-Moorhead, ND-MN
Honolulu, HI
Lake Charles, LA

11 Patents
El Paso, TX
Fayetteville-Springdale-Rogers, AR
Springfield, IL
Terre Haute, IN

12 Patents
Augusta-Aiken, GA-SC
Bakersfield, CA

Daytona Beach, FL
Glen Falls, NY
Montgomery, AL
Muncie, IN
Pensacola, FL
San Luis Obispo-Atascadero-Paso
    Robles, CA
Stockton-Lodi, CA

13 Patents
Macon, GA
Mansfield, OH

14 Patents
Fort Myers-Cape Coral, FL
Las Vegas, NV
Mobile, AL
Santa Fe, NM
Shreveport-Bossier City, LA
Waterloo-Cedar Falls, IA

15 Patents
Reno, NV
Salinas, CA
Wausau, WI

16 Patents
Lubbock, TX
Modesto, CA
Utica-Rome, NY

17 Patents
Fresno, CA
Lafayette, LA
Longview-Marshall, TX
Sherman-Denison, TX
Springfield, MO
Williamsport, PA

18 Patents
Barnstable-Yarmouth, MA
Green Bay, WI
Iowa City, IA

19 Patents
Hickory-Morganton-Lenoir, NC
Lakeland-Winter Haven, FL
Sheboygan, WI

20 Patents
Scranton-Wilkes-Barre-Hazleton, PA

21 Patents
Asheville, NC
Bloomington, IN
Janesville-Beloit, WI
Little Rock-North Little Rock, AR
Sarasota-Bradenton, FL

22 Patents
Lawrence, KS

23 Patents
Beaumont-Port Arthur, TX
Charlottesville, VA

24 Patents
Athens, GA
Benton Harbor, MI
Eugene-Springfield, OR
Portland, ME

25 Patents
Charleston-North Charleston, SC
Jackson, MI
Lima, OH
Pittsfield, MA
Tulsa, OK

26 Patents
Omaha, NE

27 Patents
Fort Pierce-Port St. Lucie, FL
Provo-Orem, UT
State College, PA
Wilmington, NC

28 Patents
Chattanooga, TN-GA
Parkersburg-Marietta, WV-OH

29 Patents
Birmingham, AL

30 Patents
Champaign-Urbana, IL

32 Patents
Bryan-College Station, TX

33 Patents
Lincoln, NE

34 Patents
Davenport-Moline-Rock Island, IA-IL

35 Patents
Des Moines, IA
Eau Claire, WI
Reading, PA

36 Patents
Elkhart-Goshen, IN

37 Patents
Elmira, NY

39 Patents
Roanoke, VA

40 Patents
Erie, PA

41 Patents
Richland-Kennewick-Pasco, WA
Wichita, KS

43 Patents
Evansville-Henderson, IN-KY

44 Patents
Huntsville, AL
Lafayette, IN

45 Patents
Columbia, SC
Norfolk-Virginia Beach-Newport
News, VA-NC

46 Patents
Lansing-East Lansing, MI
Lynchburg, VA

47 Patents
Jacksonville, FL
Youngstown-Warren, OH

48 Patents
Spokane, WA

49 Patents
Kokomo, IN

52 Patents
York, PA

53 Patents
Nashville, TN

54 Patents
Louisville, KY-IN

56 Patents
Lexington, KY

58 Patents
Appleton-Oshkosh-Neenah, WI
Springfield, MA

59 Patents
Knoxville, TN

60 Patents
Charleston, WV

61 Patents
Oklahoma City, OK
South Bend, IN

62 Patents
Gainesville, FL

68 Patents
Burlington, VT
Canton-Massillon, OH
Rochester, MN

75 Patents
Cedar Rapids, IA

76 Patents
Memphis, TN-AR-MS

77 Patents
Peoria-Pekin, IL
San Antonio, TX

80 Patents
New Orleans, LA

81 Patents
Albuquerque, NM

90 Patents
Columbus, OH
Colorado Springs, CO

93 Patents
Charlotte-Gastonia-Rock Hill, NC-
SC

94 Patents

New London-Norwich, CT

95 Patents
Melbourne-Titusville-Palm Bay, FL

96 Patents
Santa Barbara-Santa Maria-Lompoc,
CA

99 Patents
Fort Wayne, IN
Richmond-Petersburg, VA

100 Patents
Kansas City, MO-KS

103 Patents
Lancaster, PA

104 Patents
Toledo, OH

114 Patents
Fort Collins-Loveland, CO

120 Patents
Providence-Warwick-Pawtucket, RI

121 Patents
Kalamazoo-Battle Creek, MI
Sacramento-Yolo, CA

122 Patents
Johnson City-Kingsport-Bristol, TN-
VA

123 Patents
Madison, WI

124 Patents
Greenville-Spartanburg-Anderson,
SC

127 Patents
Greensboro-Winston-Salem-High
Point, NC
Orlando, FL

130 Patents
Tucson, AZ

131 Patents
Rockford, IL

133 Patents
Binghamton, NY

146 Patents
Tampa-St. Petersburg-Clearwater, FL

153 Patents
Syracuse, NY

154 Patents
Grand Rapids-Muskegon-Holland,
MI

163 Patents
Baton Rouge, LA

172 Patents
Harrisburg-Lebanon-Carlisle, PA

192 Patents
Dayton-Springfield, OH

205 Patents
Salt Lake City-Ogden, UT

211 Patents
Boise City, ID

218 Patents
Buffalo-Niagara Falls, NY

237 Patents
Portland-Salem, OR-WA

242 Patents
West Palm Beach-Boca Raton, FL

246 Patents
Allentown-Bethlehem-Easton, PA

287 Patents
Saginaw-Bay City-Midland, MI

294 Patents
Denver-Boulder-Greeley, CO

306 Patents
Miami-Fort Lauderdale, FL

332 Patents
Atlanta, GA

339 Patents
Raleigh-Durham-Chapel Hill, NC

346 Patents
Indianapolis, IN

359 Patents
Milwaukee-Racine, WI

393 Patents
St. Louis, MO

401 Patents
Hartford, CT

445 Patents
Houston-Galveston-Brazoria, TX

452 Patents
Cincinnati-Hamilton, OH

496 Patents
Seattle-Tacoma-Bremerton, WA

507 Patents
Albany-Schenectady-Troy, NY

567 Patents
Pittsburgh, PA

578 Patents
Phoenix-Mesa, AZ

607 Patents
Austin-San Marcos, TX

616 Patents
San Diego, CA

626 Patents
Washington-Baltimore, DC-MD-VA-
WV

761 Patents
Cleveland-Akron, OH

809 Patents
Dallas-Fort Worth, TX

1093 Patents
Minneapolis-St. Paul, MN-WI

1297 Patents
Detroit-Ann Arbor-Flint, MI

1384 Patents
Rochester, NY

1631 Patents
Philadelphia-Wilmington-Atlantic
City, PA-NJ-DE-MD

1929 Patents
Chicago-Gary-Kenosha, IL-IN-WI

1992 Patents
Los Angeles-Riverside-Orange
County, CA

2160 Patents
Boston-Worcester-Lawrence-Lowell-
Brockton, MA-NH

3270 Patents
San Francisco-Oakland-San Jose, CA

4674 Patents
New York-Northern New Jersey-
Long Island, NY-NJ-CT-PA

NOTES

[1] For the purpose of this study, 'innovation' refers to a commercially viable process or product developed from the appropriation of available knowledge and physical resources. Innovation, in effect, evolves from a process transforming private and public knowledge (along with physical resources) into new commercial activity.

[2] See, for example, Acs et al. (1994); Anselin et al. (1997, 2000); Audretsch and Feldman (1996a, 1996b, 1999); Jaffe (1989); Audretsch and Stephan (1996); and Jaffe et al. (1993).

[3] A small vein of related research explored R&D spillovers and social returns (Evenson 1968; Griliches 1964; Mansfield et al. 1977).

[4] The U.S. Congress established the Small Business Innovation Research Program in 1982 as a means to increase the innovative output of small firms and the share of federal R&D funds allocated to small businesses. Chapter 2 provides a detailed explanation of the SBIR Program.

[5] The 273 metropolitan areas comprise all metropolitan areas in the United States during 1990-95. Detailed information on the definition of the 273 metropolitan areas is in Chapter 4.

[6] Concern about diminishing innovation within the United States was not only felt in the political arena but also by the general population, indicated by its frequent documentation in the popular press. (See, for example, *Business Week* 1978; *The Washington Post* 1978a, 1978b, 1978c; *Time* 1978.)

[7] See, for example, Birch (1981), Data Resources, Inc. (1977), U.S. General Accounting Office (1981), Mansfield (1968), MIT Development Foundation (1975, 1979), Norris (1978), OECD (1982), Scheirer (1977), Scherer (1970), and U.S. Department of Commerce (1975).

[8] Sufficient concern among policymakers prompted joint Congressional hearings on the underutilization of small firms in federal efforts to stimulate innovation (U.S. Senate and U.S. House 1978).

[9] Federal policies targeting small business also included implementing special considerations for small firms in federal procurement policies and expansion of SBA activities, such as loan programs (U.S. Senate 1978; U.S. Small Business Administration 1979).

[10] Phase I awards were capped at $50,000 until the 1992 reauthorization.

[11] As with Phase I, Phase II awards had a lower cap ($500,000) until the 1992 reauthorization.

[12] See Wessner (2000) for an analysis of DOD's Fast Track Program.

[13] 1984 was the first year Phase II awards were given, resulting in a twelve-year period of observation.

[14] See Feldman (1994b) for a detailed breakdown of the distribution of selected innovation measures by state.

[15]R&D activity in the automobile industry largely drives Michigan's rank among the top R&D states.

[16]Utility patents are patents granted for inventions, excluding other patent types such as designs.

[17]This SBIR initiative is similar in principle to the Experimental Program to Stimulate Competitive Research (EPSCoR) Program at the National Science Foundation, which targets states with marginal sponsored research activity.

[18]Shepherd (1979 p. 400) goes so far as to assert "most of the eighty thousand patents issued each year are worthless and are never used" suggesting patents are not strongly related to innovative output.

[19]For information on the collection of these data and a critique of their use, see Feldman and Audretsch (1999). Since the late 1980s, several studies have made use of this innovation counts data to examine industrial productivity (Acs and Audretsch 1989, 1990, 1996; Acs et al. 1994; Anselin et al. 1997, 2000; Audretsch and Feldman 1996a, 1996b; Feldman 1994a, 1994b; Feldman and Florida 1994).

[20]In a broader context, this has been an issue raised regarding federal funding for U.S. science. Research budgets across many agencies have stagnated or diminished, while those at some agencies have grown rapidly. The budgetary decreases have tended to occur at agencies targeting the physical sciences, and agencies, such as NIH, that target biological sciences have experienced significant increases. This shift in funding can dramatically influence the magnitude of research performed across fields of science.

[21]Contrarily, this effect may lead to the opposite outcome. Areas with low innovation but growing SBIR activity may begin to experience a rise in innovative activity as a consequence of SBIR research.

[22]For an overview of the relationship between agglomeration and innovation in the high-technology sector, see Oakey (1984).

[23]Knowledge spillovers will be discussed in detail in the next section of this chapter.

[24]It can also result in the increased production of skilled workers in the geographic area as educational institutions respond to the demand for specific kinds of skilled labor in the area.

[25]See Stephan (1996) for an overview of literature related to knowledge spillovers and productivity.

[26]It could be argued that the ease with which universities disseminate knowledge is lessening due to increasing intellectual property protection and commercialization strategies at universities.

[27]The data on actual innovations were collected for 1982 only, as a census of innovation citations conducted by the Small Business Administration; no publicly available data on more recent innovation counts exist.

[28]See Chapter 3 for a discussion of the knowledge production function and the local technological infrastructure.

[29]The 273 metro areas are comprised of 245 Metropolitan Statistical Areas (MSAs), 17 Consolidated Metropolitan Statistical Areas (CMSAs), and 11 New England County Metropolitan Areas (NECMAs). During the 1990-95 period, the United States had 332 classified metropolitan areas. This analysis excludes 59 of the 332 metro areas (all 58 Primary Metropolitan Statistical Areas (PMSAs) and 1 NECMA that serves as a PMSA substitute) because these areas are located within the 17 CMSAs.

[30]In the analysis, the grouping of industries at the two-digit SIC code level reduces the measurement error from this method, given the broad scope of business activities at such an aggregated level.

[31]The *Directory of American Research and Technology* is available in print version only, requiring extracted data to be encoded by hand.

[32]The statistical significance of the difference in the means was tested using the Approximate t Statistic, which assumes unequal variances across groups. This statistic was selected after rejecting the hypothesis that the variances across the two groups are equal (Satterthwaite 1946; Steel and Torrie 1980).

[33]Hausman et al. (1984) provide the first application of count model estimation to innovative activity when investigating the effects of industrial R&D on firms' patenting behavior using panel data.

[34]All tests for statistical significance in this analysis are based on a two-tailed t-test at the 1, 5, and 10 percent levels of significance.

[35]The hurdle model was also estimated using the level of population instead of population density. Little difference emerges between the results for population and population density. The population results reinforce the implication that area size is insignificant for Phase II activity and, therefore, are unreported.

[36]Negative selection bias is also indicated in an unreported estimation of the model using nonlinear least squares following Terza (1998).

[37]This supports previous research documenting strong ties between university research and the electronics industry (Mansfield 1991a, 1991b, 1995; Saxenian 1985).

[38]Appendix D reports empirical results for the two-step hurdle model across the top three agencies but not by individual industry. The five industries are aggregated to comprise a 'high-tech' sector. These results indicate significant spillovers from R&D labs and universities on the likelihood of Phase II activity for all three agencies. The negative binomial equations suggest that university R&D activity has a significant impact on the number of Phase II awards for awards funded by HHS and NASA but not those funded by DOD. The spillovers from R&D labs present in the probit equations are no longer evident. However, a significant and negative effect of the size of the metropolitan area on the number of Phase II awards funded by all three agencies is present. This suggests that increases in population density are associated with declines in agency-level Phase II activity, implying agglomerative diseconomies due to area size are present.

[39]These results, however, are reported in Table 32 of Appendix D.

[40]This in itself suggests a strong relationship between the presence of research universities and the occurrence of Phase II activity in a metropolitan area. In the machinery industry during 1990-95, every metropolitan area with some level of Phase II activity also had at least one research university. Appendix D reports estimation results for the probit equation by agency and industry where the specification replaces UNIVDUM with UNIVR&D, which allows estimation of the equation to occur for industrial machinery. Interestingly, the only significant effect in the HHS equation is for R&D labs—the only time R&D labs has a significant impact on the likelihood of Phase II activity funded by HHS across all five industries.

[41]For a discussion of the independent variables, see Chapter 4.

[42]Several alternative approaches have been pursued to counter some of the drawbacks of the USPTO concordance. Evenson has been instrumental in developing the Yale Technology Concordance and Wellesley Technology Concordance, which are based on patent examiners' assignment of industry classifications in the Canadian patent system (Englander et al. 1988; Evenson et al. 1988; Johnson 1999; Johnson and Evenson 1997; Kortum and Putnam 1989, 1997). Scherer (1965a, 1965b) and the NBER group (Bound et al. 1984) have compiled industry-level patent data by aggregating firm level data based on the firms' primary activities. Others (Griliches et al. 1987; Hall et al. 1988; Scherer 1982b, 1984) have linked individual patent data with other data sources, including the Federal Trade Commission Line of Business survey and financial data for publicly traded firms.

[43]The unconditional model would be based on all metropolitan areas regardless of their patent activity. The conditional model for patent counts is based only on metropolitan areas with positive patent activity.

[44]The unreported patent model controlling for selection is based on a probit equation with one independent variable. The probit equation models the likelihood of patenting as depending only on the presence of research activity in the metropolitan area. This research activity variable is defined as a dummy variable indicating whether (=1) or not (=0) a metropolitan area had either R&D labs or research universities located in the area. The negative binomial equation is the same as in the model reported in this chapter.

[45]As with the SBIR Phase II data, overdispersion occurs in the patent counts, evidenced by the significance of the overdispersion parameter, α. This coincides with previous evidence of overdispersion in patent data (Adams 1998). The presence of overdispersion indicates the appropriateness of the negative binomial model, as was the case with Phase II awards.

[46]Recent changes to the structure of the SBIR Program also need to be taken into account in future research that examines SBIR activity after 1995. For instance, the mandatory set aside for SBIR research increased in 1997, which likely affected the number of SBIR awards.

REFERENCES

Acs, Zoltan. 1999. *Are Small Firms Important? Their Role and Impact.* Boston: Kluwer Academic Publishers.

Acs, Zoltan and David Audretsch. 1989. Patents as a Measure of Innovative Activity. *KKKLOS* 42: 171-80.

_____. 1990. *Innovation and Small Firms.* Cambridge, MA: The MIT Press.

_____. 1993. Has the Role of Small Firms Changed in the United States? In *Small Firms and Entrepreneurship: an East-West Perspective*, eds. Zoltan Acs and David Audretsch, 55-77. Cambridge, UK: Cambridge University Press.

_____. 1996. Innovation in Large and Small Firms: an Empirical Analysis. In *Small Firms and Economic Growth*, Vol. 1, Cheltenham, UK: Elgar: 393-405.

Acs, Zoltan, David Audretsch and Maryann Feldman. 1992. Real Effects of Academic Research: Comment. *American Economic Review* 82, No. 1: 363-367.

_____. 1994. R&D Spillovers and Recipient Firm Size. *The Review of Economics and Statistics* 100, No. 2: 336-40.

Adams, James. 1998. Endogenous R&D Spillovers and Industrial Research Productivity. University of Florida Working Paper (December).

_____. 2001. Comparative Localization of Academic and Industrial Spillovers. National Bureau of Economic Research Working Paper 8292 (May).

Almeida, Paul and Bruce Kogut. 1998. The Exploration of Technological Diversity and the Geographic Localization of Innovation. *Small Business Economics* 9: 21-31.

Angel, David. 1991. High-Technology Agglomeration and the Labor Market: The Case of Silicon Valley. *Environment and Planning* A 23, No. 10 (October): 1501-16.

Anselin, Luc, Attila Varga and Zoltan Acs. 1997. Local Geographic Spillovers Between University Research and High Technology Innovations. *Journal of Urban Economics* 42: 422-48.

_____. 2000. Geographic and Sectoral Characteristics of Academic Knowledge Externalities. *Papers in Regional Science*, 79, No. 4: 435-443.

Archibald, Robert and David Finifter. 2000. Evaluation of the Department of Defense Small Business Innovation Research Program and Fast Track Initiative: A Balanced Approach. In *The Small Business Innovation Research Program: An Assessment of the Department of Defense Fast Track Initiative*, ed. Charles Wessner. Washington, DC: National Academy Press.

Arrow, Kenneth. 1962. Economic Welfare and the Allocation of Resources for Invention. In *The Rate and Direction of Inventive Activity*, ed. Richard Nelson. Princeton, NJ: Princeton University Press.

Audretsch, David and Maryann Feldman. 1996a. Innovative Clusters and the Industry Life Cycle. *Review of Industrial Organization* 11, No. 2: 253-73.

_____. 1996b. R&D Spillovers and the Geography of Innovation and Production. *American Economic Review* 86, No. 3: 630-40.

Audretsch, David, Albert Link, and John Scott. 2000. Statistical Analysis of the National Academy of Sciences Survey of Small Business Innovation Research Awardees: Analyzing the Influence of the Fast Track Program. In *The Small Business Innovation Research Program: An Assessment of the Department of Defense Fast Track Initiative,* ed. Charles Wessner. Washington, DC: National Academy Press.

Audretsch, David and Paula Stephan. 1996. Company-Scientist Locational Links: the Case of Biotechnology. *American Economic Review* 86, No.3: 641-652.

_____. 1999. How and Why Does Knowledge Spill Over in Biotechnology? In *Innovation, Industry Evolution, and Employment*, eds. David Audretsch and Roy Thurik, 216-229. Cambridge, UK: Cambridge University Press.

Bania, Neil, Randall Eberts and Michael Fogarty. 1993. Universities and the Startup of New Companies: Can We Generalize from Route 128 and Silicon Valley? *The Review of Economics and Statistics* 65 (August): 761-766.

Bartel, Ann and Frank Lichtenberg. 1987. The Comparative Advantage of Educated Workers in Implementing New Technology. *Review of Economics and Statistics* 69, No. 1(February): 1-11.

Beeson, Patricia. 1992. Agglomeration Economies and Productivity Growth. In *Sources of Metropolitan Growth*, eds. Edwin Mills and John McDonald, 19-35. New Brunswick, NJ: Center for Urban Policy Research.

Birch, David. 1981. Who Creates Jobs? *The Public Interest* 65 (fall): 3-14.

Bound, John, Clint Cummins, Zvi Griliches, Bronwyn Hall, and Adam Jaffe. 1984. Who Does R&D and Who Patents? In *R&D, Patents, and Productivity*, ed. Zvi Griliches. Chicago: University of Chicago Press.

Bunk, Steve. 1999. How to Manage Knowledge and "Gold Collar" Workers. *The Scientist* (February): 16-17.

Busch, Chris. Prepared Testimony of Dr. Chris W. Busch, SBIR Consultant, Ronan, Montana. In *The Small Business Innovation Research (SBIR) Program* Hearings, U.S. House, Small Business Committee, 27 May. Washington, DC: U.S. Government Printing Office.

Business Week. 1978. Vanishing Innovation (3 July).

Cahill, Peter. 2000. Fast Track: Is It Speeding Commercialization of the Department of Defense Small Business Innovation Research Projects? In *The Small Business Innovation Research Program: An Assessment of the Department of Defense Fast Track Initiative*, ed. Charles Wessner. Washington, DC: National Academy Press.

Cameron, Adrian and Pravin Trivedi. 1998. *Regression Analysis of Count Data*. Cambridge, UK: Cambridge University Press.

Cardenas, Michael. 1981. Statement of Michael Cardenas, Administrator, U.S. Small Business Administration, Before the Subcommittee on Innovation and Technology, Committee on Small Business, United States Senate, July 16, 1981. In *The Small Business Innovation Research Act of 1981* Hearings, U.S. Senate. Washington, DC: U.S. Government Printing Office.

Carlsson, Bo and Pontus Braunerhjelm. 1999. Industry Clusters: Biotechnology/ Biomedicine and Polymers in Ohio and Sweden. In *Innovation, Industry Evolution, and Employment*, eds. David Audretsch and Roy Thurik, 182-215. Cambridge, UK: Cambridge University Press.

Carlsson, Bo and Rikard Stankiewicz. 1991. On the Nature, Function, and Composition of Technological Systems. *Journal of Evolutionary Economics* 1, No. 2: 93-118.

Carter, Jimmy. 1979. *Message From the President of the United States Transmitting A Science and Technology Policy for the Future*. House Document No. 96-81. Washington, DC: US Government Printing Office.

Coffey, William and Mario Polese. 1987. Trade and Location of Producer Services: A Canadian Perspective. *Environment and Planning* A 19, No. 5 (May): 597-611.

Cohen, Wesley and Richard Levin. 1989. Empirical Studies of Innovation and Market Structure. In *Handbook of Industrial Organization*, vol. 2, eds. Richard Schmalensee and Robert Willig, 1059-1107. Amsterdam, The Netherlands: Elsevier Science Publishers.

Data Resources, Inc. 1977. The Role of High-Technology Industries in Economic Growth. Prepared for General Electric Company (March 15).

DeVol, Ross. 1999. *America's High-Tech Economy*. Santa Monica, CA: Milken Institute.

Dorfman, Nancy. 1983. Route 128: the Development of a Regional High Technology Economy. *Research Policy* 12: 299-316.

Dumais, Guy, Glenn Ellison, and Edward Glaeser. 1997. Geographic Concentration as a Dynamic Process. National Bureau of Economic Research Working Paper 6270.

Engert, Clyde. 1998. An Experimental Cooperative SSBIR/SSTTR Program Leading to Self-Funding. Final Report for the SSBIR. Kansas City: Kansas Technology Enterprise Corporation.\

Englander, A. Steven, Robert Evenson, and Masaharu Hanazaki. 1998. R&D, Innovation, and the Total Factor Productivity Slowdown. *OECD Economic Studies* 0, No. 11 (autumn): 8-42.

Enos, John. 1962. *Invention and Innovation in the Petroleum Refining Industry*, IA: Nelson.

Evenson, Robert. 1968. The Contributions of Agricultural Research and Extension to Agricultural Production. Ph.D. dissertation, University of Chicago.

Evenson, Robert, Samuel Kortum, and Jonathan Putnam. 1988. Estimating Patents by Industry Using the Yale-Canada Patent Concordance. Unpublished manuscript, Yale University.

Feldman, Maryann. 1994a. Knowledge Complementarity and Innovation. *Small Business Economics* 6, No. 5 (October): 363-72.

_____. 1994b. *The Geography of Innovation*. Dordrecht, The Netherlands: Kluwer Academic Publishers.

Feldman, Maryann and David Audretsch. 1999. Innovation in Cities: Science-based Diversity, Specialization and Localized Competition. *European Economic Review* 43: 409-429.

Feldman, Maryann and Richard Florida. 1994. The Geographic Sources of Innovation: Technological Infrastructures and Product Innovation in the United States. *Annals of the Association of American Geographers* 84: 210-229.

Finch, Brian. 1987. Defense Spending and Regional Growth. Master's thesis, Naval Postgraduate School.

Flender, John and Richard Morse. 1975. The Role of New Technical Enterprises in the U.S. Economy. Cambridge, MA: MIT Development Foundation (1 October).

Fosler, Scott, ed. 1988. *The New Economic Role of American States: Strategies in a Competitive World Economy*. Oxford: Oxford University Press.

Fritsch, Michael and Rolf Lukas. 1999. Innovation, Cooperation, and the Region. In *Innovation, Industry Evolution, and Employment*, eds. David Audretsch and Roy Thurik, 157-181. Cambridge, UK: Cambridge University Press.

Gans, Joshua and Scott Stern. 2000. When Does Funding Research by Smaller Firms Bear Fruit? Evidence from the SBIR Program. National Bureau of Economic Research Working Paper 7877 (September).

Gansler, Jacques. 1980. *The Defense Industry*. Cambridge, MA: MIT Press.

Gellman Research Associates. 1976. Firm Size and Technological Innovation. Report prepared for the U.S. Small Business Administration. Washington, DC: Gellman Research Associates.

Glaeser, Edward. 2000. Urban and Regional Growth. In *The Handbook of Economic Geography*, eds. Gordon Clark, Maryann Feldman, and Meric Gertler, 83-98. Oxford: Oxford University Press.

Glasmeier, Amy. 1986. High-Tech Industries and the Regional Division of Labor. *Industrial Relations* 25, No. 2 (spring): 197-211.

_____. 1988. Factors Governing the Development of High Tech Industry Agglomerations: a Tale of Three Cities. *Regional Studies* 22, No. 4 (August): 287-301.

Greene, William. 1993. *Econometric Analysis*. Englewood Cliffs, NJ: Prentice Hall.

_____. 1994. Accounting for Excess Zeros and Sample Selection in Poisson and Negative Binomial Regression Models. Working Paper Number 94-10, Department of Economics, Stern School of Business, New York University.

Griliches, Zvi. 1964. Research Expenditures, Education, and the Aggregate Agricultural Production Function. *American Economic Review* 54, No. 6: 961-974.

_____. 1979. Issues in Assessing the Contribution of Research and Development to Productivity Growth. *Bell Journal of Economics* 10, No. 1 (spring): 92-116.
_____. 1990. Patent Statistics as Economic Indicators: A Survey. *Journal of Economic Literature* 28 (December): 1661-1707.

_____. 1992. The Search for R&D Spillovers. National Bureau of Economic Research Working Paper 3768.

Griliches, Zvi, Ariel Pakes, and Bronwyn Hall. 1987. The Value of Patents as Indicators of Inventive Activity. In *Economic Policy and Technological Performance*, eds. Partha Dasgupta and Paul Stoneman, 97-124. Cambridge, UK: Cambridge University Press.

Hadlock, Paul, Daniel Hecker, and Joseph Gannon. 1991. High Technology Employment: Another View. *Monthly Labor Review* 114, No. 7 (July): 26-31.

Hall, Bronwyn, Clint Cummins, Elizabeth Laderman, and Joy Mundy. 1988. The R&D Master File. National Bureau of Economic Research Technical Paper 72.

Hamberg, David. 1963. Invention in the Industrial Laboratory. *Journal of Political Economy* (April).

Hausman, Jerry, Bronwyn Hall, and Zvi Griliches. 1984. Econometric Models for Count Data with an Application to the Patents-R&D Relationship. *Econometrica* 52, No. 4 (July): 909-38.

Hecker, Daniel. 1999. High Technology Employment: A Broader View. *Monthly Labor Review* 122, No. 6 (June): 18-28.

Heckman, James. 1979. Sample Selection Bias as a Specification Error. *Econometrica* 47, No. 1 (January): 153-61.

Henderson, Rebecca and Iain Cockburn. 1996. Scale, Scope, and Spillovers: The Determinants of Research Productivity in Drug Discovery. *Rand Journal of Economics* 27, No. 1 (spring): 32-59.

Hubbard, Glenn. 1998. Capital-market Imperfection and Investment. *Journal of Economic Literature* 36: 193-225.

Huffman, Wallace and Robert Evenson. 1991. *Science for Agriculture*. Ames, IA: Iowa State University Press.

Jacobs, Jane. 1960. *The Economy of Cities*. New York, NY: Random House.

Jaffe, Adam. 1986. Technological Opportunity and Spillovers of R&D: Evidence from Firms' Patents, Profits, and Market Value. *American Economic Review* 76 (December): 984-1001.

_____. 1989. Real Effects of Academic Research. *American Economic Review* 79 (December): 957-70.

Jaffe, Adam, Manuel Trajtenberg and Rebecca Henderson. 1993. Geographic Localization of Knowledge Spillovers as Evidenced by Patent Citations. *The Quarterly Journal of Economics* (August): 577-98.

Jewkes, John, David Sawyers and Richard Stillerman. 1969. *The Sources of Invention*. 2nd ed. New York: W.W. Norton.

Johnson, Daniel. 1999. 150 Years of American Invention: Methodology and a First Geographical Application. Wellesley College Working Paper 99-01 (January).

Johnson, Daniel and Robert Evenson. 1997. Innovation and Invention in Canada. *Economic Systems Research* 9, No. 2: 177-192.

Keefe, Bob. 2001. Biotech Industry Newest Darling. *The Atlanta Journal-Constitution* (27 June): E1, E3.

Korobow, Adam. 2002. *Entrepreneurial Wage Dynamics in the Knowledge Economy*. Dordrecht, The Netherlands: Kluwer Academic Publishers.

Kortum, Samuel and Jonathan Putnam. 1989. Estimating Patents by Industry: Parts I and II. Papers presented at the National Bureau of Economic Research Conference on Patent Count and Patent Renewal Data (August).

_____. 1997. Assigning Patents to Industries: Tests of the Yale Technology Concordance. *Economic Systems Research* 9, No. 2: 161-175.

Krugman, Paul. 1991a. Increasing Returns and Economic Geography. *Journal of Political Economy* 99, No. 3 (June): 483-99.

_____. 1991b. *Geography and Trade*. Cambridge, MA: MIT Press.

Lucas, Richard. 1993. Making a Miracle. *Econometrica* 61, No. 2 (March): 251-72.

Lucas, Robert. 1988. On the Mechanics of Economic Development. *Journal of Monetary Economics* 22, No. 1 (July): 3-42.

Luker, William and Donald Lyons. 1997. Employment Shifts in High-technology Industries, 1988-96. *Monthly Labor Review* 120, No. 6 (June): 12-25.

MacPherson, Alan. 1991. Interfirm Information Linkages in an Economically Disadvantaged Region: An Empirical Perspective from Metropolitan Buffalo. *Environment and Planning A*: 591-606.

Malecki, Edward. 1983. Technology and Regional Development: a Survey. *International Regional Science Review* 8: 89-125.

_____. 1985. Public Sector Research and Development and Regional Economic Performance in the United States. In *The Regional Economic Impact of Technological Change*, eds. A. Thwaites and R. Oakey. London: Pinter.

_____. 1991. *Technology and Economic Development: the Dynamics of Local, Regional, and National Change*. Essex, UK: Longman Scientific & Technical.

Malecki, Edward and Susan Bradbury. 1992. R&D Facilities and Professional Labour: Labour Force Dynamics in Technology. *Regional Studies* 26, No. 2: 123-36.

Mamuneas, Thefanis. 1999. Spillovers from Publicly Financed R&D Capital in High-Tech Industries. *International Journal of Industrial Organization* 17, No. 2 (February): 215-39.

Mansfield, Edwin. 1968. *Industrial Research and Technological Innovation*. W.W. Norton.

_____. 1984. Comment on Using Linked Patent and R&D Data to Measure Interindustry Technology Flows. In *R&D, Patents, and Productivity*, ed. Zvi Griliches, 462-464. Chicago: University of Chicago Press.

_____. 1991a. Academic Research and Industrial Innovation. *Research Policy* 20: 1-12.

_____. 1991b. Estimates of the Social Returns from Research and Development. In *Science and Technology Yearbook*, eds. M. Meredith, S. Nelson, and A. Teich, 313-320. Washington, DC: American Association for the Advancement of Science.

_____. 1995. Academic Research Underlying Industrial Innovations: Sources, Characteristics, and Financing. *The Review of Economics and Statistics* 77, No. 1 (February): 55-65.

Mansfield, Edwin, John Rapoport, Anthony Romero, Samuel Wagner, and George Beardsley. 1977. Social and Private Rates of Returns from Industrial Innovations. *Quarterly Journal of Economics* 91, No. 2: 221-240.

Manski, Charles. 1995. *Identification Problems in the Social Sciences*. Cambridge, MA: Harvard University Press.

Markusen, Ann. 1986. Defense Spending and the Geography of High-Tech Industries. In *Technology, Regions, and Policy*, ed. J. Rees, 94-119. New Jersey: Rowman and Littlefield.

Markusen, Ann, Peter Hall, and Amy Glasmeier. 1986. *High Tech America: The What, How, Where, and Why of the Sunrise Industries*. Boston: Allen & Unwin.

Martinez, Nieves and Pedro Nueño. 1988. Catalan Regional Business Culture, Entrepreneurship, and Management Behavior: an Exploratory Study. In *Regional Cultures, Managerial Behavior, and Entrepreneurship: An International Perspective*, ed. J. Weiss, 61-75. New York: Quorum Books.

Mazza, J. and D. Wilkinson. 1980. The Unprotected Flank: Regional and Strategic Imbalances in Defense Spending Patterns. Washington, DC: Northeast-Midwest Institute.

Mehay, Stephen and Loren Solnick. 1990. Defense Spending and State Economic Growth. *Journal of Regional Science* 30, No. 4: 477-487.

MIT Development Foundation. 1979. The Job Generating Process. Cambridge, MA: MIT Development Foundation.

Mullahy, John. 1986. Specification and Testing in Some Modified Count Data Models. *Journal of Econometrics* 33: 341-365.

Murphy, Kevin and Robert Topel. 1985. Estimation and Inference in Two-step Econometric Models. *Journal of Business and Economic Statistics* 3: 370-79.

Narin, Francis, Kimberly Hamilton, and Dominic Olivastro. 1997. The Increasing Linkage Between U.S. Technology and Public Science. *Research Policy* 26, No. 3 (October): 317-330.

National Academy of Sciences. 1999. Harnessing Science and Technology for America's Economic Future. Washington, DC: National Academy Press.

National Science Board. 1977. Science and Engineering Indicators 1976. Washington, DC: U.S. Government Printing Office.

_____. 1998. *Science and Engineering Indicators 1998*. Arlington, VA: U.S. Government Printing Office: 6-2-6-16.

_____. 2000. *Science and Engineering Indicators 2000*. Washington, DC: U.S. Government Printing Office.

Norris, William. 1978. Recommendations for Creating Jobs Through the Success of Small Innovative Businesses. A Report to the Assistant Secretary of Commerce for Science and Technology. Washington, DC: Control Data, Inc. (December).

Oakey, Ray. 1984. *High Technology Small Firms: Regional Development in Britain and the United State*. New York, NY: St. Martin's Press.

Organization for Economic Cooperation and Development (OECD). 1982. *Innovation in Small and Medium Firms*. Paris: OECD.

Office of Management and Budget. 1987. *Standard Industrial Classification Manual 1987*, Springfield, VA: National Technical Information Service.

Office of Research. 2000. 'Teaming to Win:' Wyoming SBIR/STTR Initiative Goals and Strategies for the Years 2001 and Beyond. Laramie, WY: University of Wyoming (October 9).

O'Sullivan, Arthur. 2000. *Urban Economics*, 4th ed., Irwin McGraw-Hill.

Pavitt, Keith, Michael Robson and Joe Townsend. 1987. The Size Distribution of Innovating Firms in the UK: 1945-1983. *The Journal of Industrial Economics* 55: 291-316.

Phillips, Bruce. 1991. The Increasing Role of Small Firms in the High-Technology Sector: Evidence from the 1980s. *Business Economics* 26, No. 1 (January): 40-47.

Pohlmeier, Winfried and Volker Ulrich. 1992. Contact Decisions and Frequency Decision: An Econometric Model of the Demand for Ambulatory Services. ZEW Discussion Paper 92-09.

Rauch, James. 1993. Productivity Gains from Geographic Concentration of Human Capital: Evidence from the Cities. *Journal of Urban Economics* 34: 380-400.

Rees, John. 1981. The Impact of Defense Spending on Regional Industrial Change in the United States. In *Federalism and Regional Development*, ed. G. Hoffman. Austin: University of Texas Press.

_____. 1982. Defense Spending and Regional Industrial Change. *Texas Business Review* (January-February): 40-44.

Rees, John and Howard Stafford. 1986. Theories of Regional Growth and Industrial Location: Their Relevance for Understanding High-Technology Complexes. In *Technology, Regions, and Policy*, ed. J. Rees, 23-50. New Jersey: Rowman and Littlefield.

Romer, Paul. 1986. Increasing Returns and Long-run Growth. *Journal of Political Economy* 64: 1002-37.

_____. 1990. Endogenous Technological Change. *Journal of Political Economy* 98, No. 5 Part 2 (October): S71-S102.

Rosenthal, Stuart and William Strange. 2001. The Determinants of Agglomeration. *Journal of Urban Economics* 50, No. 2 (September): 191-229.

Ruppert, Sandra. 2001. Where We Go From Here: State Legislative Views on Higher Education in the New Millennium. A Report of the 2001 Higher Education Issues Survey prepared for the National Education Association of the United States.

Satterthwaite, F. 1946. An Approximate Distribution of Estimates of Variance Components. *Biometrics Bulletin* 2: 110-14.

Satterthwaite, Mark. 1992. High-Growth Industries and Uneven Metropolitan Growth. In *Sources of Metropolitan Growth*, eds. Edwin Mills and John McDonald, 39-50. New Brunswick, NJ: Center for Urban Policy Research.

Saxenian, AnnaLee. 1985. Silicon Valley and Route 128: Regional Prototypes or Historical Exceptions? In *High-technology, Space and Society*, ed. M. Castells. Beverly Hills, CA: Sage.

_____. 1996. *Regional Advantage: Culture and Competition in Silicon Valley and Route 128.* Cambridge, MA: Harvard University Press.

Scheirer, William. 1977. Small Firms and Federal R&D. Prepared for the Office of Management and Budget, Washington, DC: U.S. Government Printing Office.

Scherer, Frank. 1965a. Corporate Inventive Output, Profits, and Growth. *Journal of Political Economy* 73, No. 3 (June): 290-97.

_____. 1965b. Firm Size, Market Structure, Opportunity, and the Output of Patented Inventions. *American Economic Review* 55 (December): 1097-1125.

_____. 1970. *Industrial Market Structure and Economic Performance*, Rand McNally.

_____. 1982a. The Office of Technology Assessment and Forecast Industry Concordance as a Means of Identifying Industry Technology Origins. *World Patent Information* 4, No. 1: 12-17.

_____. 1982b. Demand-pull and Technological Invention: Schmookler Revisited. *Journal of Industrial Economics* 30, No. 3 (March): 226-237.

_____. 1983. The Propensity to Patent. *International Journal of Industrial Organization* 1: 107-28.

_____. 1984. Using Linked Patent and R&D Data to Measure Interindustry Technology Flows. In *R&D, Patents, and Productivity*, ed. Zvi Griliches, 417-461. Chicago: University of Chicago Press.

Schumpeter, Joseph. 1942. *Capitalism, Socialism and Democracy.* New York, NY: Harper and Row.

Shepherd, William. 1979. *The Economics of Industrial Organization*, Englewood Cliffs, NJ: Prentice Hall.

Smilor, Raymond, George Kozmetsky and David Gibson, eds. 1988. *Creating the Technopolis: Linking Technology Commercialization and Economic Development.* Cambridge, MA: Ballinger Publishing Company.

Smith, Keith. 1997. Economic Infrastructures and Innovation Systems. In *Systems of Innovation: Technologies, Institutions and Organizations*, ed. Charles Edquist, 86-106. London: Pinter.

Soete, Luc. 1983. Comments on the OTAF Concordance Between the US SIC and the US Patent Classification. Mimeograph prepared by SPRU, University of Sussex (November).

Southern Growth Policies Board. 2001. Conference to Focus on Building Tech-Based Economies. *Friday Facts* 13, No. 28 (20 July 20).

State Science & Technology Institute. 2000. Senate SBIR Language Offers $10 Million to States. *SSTI Weekly Digest*, <http://www.ssti.org/Digest/sbirart.htm>.

Steel, Robert and James Torrie. 1980. *Principles and Procedures of Statistics.* 2nd ed. New York: McGraw-Hill.

Stephan, Paula. 1996. The Economics of Science. *Journal of Economic Literature* 34 (September): 1199-1235.

Stiglitz, Joseph and Andrew Weiss. 1981. Credit Rationing in Markets with Incomplete Information. *American Economic Review* 71: 393-409.

Storper, Michael. 1982. The Spatial Division of Labor: Technology, the Labor Process, and the Location of Industries. Ph.D. dissertation, University of California, Berkeley.

Terza, Joseph. 1998. Estimating Count Data Models with Endogenous Switching: Sample Selection and Endogenous Treatment Effects. *Journal of Econometrics* 84: 129-54.

The Illinois Coalition. 2000. The Use and Impact of the Small Business Innovation Research Program in Illinois, FY 1983-1995. Chicago: The Illinois Coalition.

The Washington Post. 1978a. Something's Happened to Yankee Ingenuity (3 September): G1.

_____. 1978b. U.S. Seen Losing Technological Edge in Some Industries (24 November): A14.

_____. 1978c. Mother of Invention, Is This the End? (25 November): A14.

Tibbetts, Roland. 1996. The Small Business Innovation Research Program and NSF SBIR Commercialization Results. Unpublished manuscript prepared for the National Science Foundation (November).

_____. 1998. An Analysis of the Distribution of SBIR Awards by States, 1983-1996. Prepared for the U.S. Small Business Administration, Office of Advocacy (January).

Time. 1978. The Innovation Recession (2 October): 57.

U.S. Congress. 1982. Public Law 97-219. *Small Business Innovation Development Act of 1982.* Washington DC, U.S. Government Printing Office, 22 July.

_____. 1992. Public Law 102-564. *Small Business Innovation Research Program Reauthorization Act of 1992.* Washington, DC: U.S. Government Printing Office, 28 October.

_____. Senate Committee on Small Business. 1981. *Small Business Research Act of 1981.* Hearings before the Senate Committee on Small Business. Washington, DC: U.S. Government Printing Office.

U.S. Department of Commerce. 1975. *The Role of New Technical Enterprises in the U.S. Economy.* Report of the Commerce Technical Advisory Board to the Secretary of Commerce, Washington, DC: U.S. Government Printing Office, 1 October.

_____. 1977. Assistant Secretary for Science and Technology. *U.S. Technology Policy.* Washington, DC: U.S. Government Printing Office.

U.S. General Accounting Office (GAO). 1981. *Consistent Criteria are Needed to Assess Small-Business Innovation Initiatives.* Report to the Congress of the United States. Washington, DC: U.S. Government Printing Office, 7 July.

_____. 1998. *Federal Research: Observations on the Small Business Innovation Research Program.* GAO/RCED-98-132 (April). Washington, DC: U.S. Government Printing Office.

_____. 1999. *Federal Research: Evaluation of Small Business Innovation Research can be Strengthened.* Report to the House of Representatives Committee on Science. Washington, DC: U.S. Government Printing Office, June.

U.S. Office of Management and Budget. Office of Federal Procurement Policy. 1977. *Small Firms and Federal Research and Development.* Washington, DC: U.S. Government Printing Office, 24 February.

U.S. Patent and Trademark Office (USPTO). 2000. United States Patent Grants by State, County, and Metropolitan Area (Utility Patents 1990-1999). Report of Technology and Forecast Assessment, April.

U.S. Senate. Select Committee on Small Business. 1978. *A Bill to Amend the Small Business Act to Expand and Revise the Procedures for Insuring Small Business Participation in Government Procurement Activities.* Hearings before the Senate Select Committee on Small Business. Washington, DC: U.S. Government Printing Office, 7 February and 18 April.

U.S. Senate and U.S. House. Senate Select Committee on Small Business and House Committee on Small Business. 1978. *Underutilization of Small Business in the Nation's Efforts to Encourage Industrial Innovation.* Hearings before the Senate Select Committee on Small Business and the House Committee on Small Business. Washington, DC: U.S. Government Printing Office, 9-10 August.

U.S. Small Business Administration (SBA). 1999. Notice of Intent to Make Funds Available. <http://www.sba.gov/sbir/intent.html> [18 January, 2001].

_____. Office of Advocacy. 1995. *Results of Three-year Commercialization Study of the SBIR Program.* Washington, DC: US Government Printing Office.

_____. Office of Advocacy. 2000. *Developing High Technology Communities: San Diego*, April.

_____. Office of the Chief Counsel for Advocacy. 1979. *Small Business and Innovation: A Report of an SBA Office of Advocacy Task Force.* Washington, DC: U.S. Government Printing Office, May.

von Hippel, Eric. 1988. *The Sources of Innovation.* New York: Oxford University Press.

_____. 1994. Sticky Information and the Locus of Problem Solving: Implications for Innovation. *Management Science* 40: 429-439.

Walcott, Susan. 1999. High Tech in the Deep South: Biomedical Firm Clusters in Metropolitan Atlanta. *Growth and Change* (winter): 48-74.

_____. 2000. *Defining and Measuring High Technology in Georgia.* Report of the Fiscal Research Program, No. 50. Atlanta, GA: Georgia State University, Fiscal Research Program (December).

Weber, Alfred. 1929. *Theory of the Location of Industries.* Chicago: University of Chicago Press.

Wessner, Charles, ed. 2000. *The Small Business Innovation Research Program: An Assessment of the Department of Defense Fast Track Initiative.* Washington, DC: National Academy Press.

Weston, D. and Philip Gummett. 1987. The Economic Impact of Military R&D: Hypotheses, Evidence, and Verification. *Defense Analysis* 3: 63-76.

Wood, P. 1969. Industrial Location and Linkage. *Area* 2: 32-39.

Wozniak, Gregory. 1984. The Adoption of Interrelated Innovations: A Human Capital Approach. *Review of Economics and Statistics* 66, No. 1 (February): 70-79.

_____. 1987. Human Capital, Information, and the Early Adoption of New Technology. *Journal of Human Resources* 22, No. 1 (winter): 101-112.

Zerbe, Richard. 1976. Research and Development by Smaller Firms. *Journal of Contemporary Business* (spring): 91-113.

Zucker, Lynne, Michael Darby, and Marilynn Brewer. 1994. Intellectual Capital and the Firm: The Technology of Geographically Localized Knowledge Spillovers. National Bureau of Economic Research Working Paper 4946.

INDEX